Salmon Aquaculture

Edited by Qian Lu

Published in London, United Kingdom

IntechOpen

Supporting open minds since 2005

Salmon Aquaculture
http://dx.doi.org/10.5772/intechopen.95211
Edited by Qian Lu

Contributors
Renzo Pepe-Victoriano, Héctor Aravena-Ambrosetti, Piera Pepe-Vargas, James E. Barasa, Purity Nasimiyu Nasimiyu Mukhongo, Cynthia Chepkemoi Chepkemoi Ngetich, Paul Gümpel, Arnulf Hörtnagl, Torsten Bogatzky, Urs Dornbierer, Jose A. Gallardo, Margarita Rivera, Cristian Araneda, Anti Vasemägi, Alexander Dindial, Aslihan Katip

First published in London, United Kingdom, 2022 by IntechOpen
IntechOpen is the global imprint of INTECHOPEN LIMITED, registered in England and Wales, registration number: 11086078, 5 Princes Gate Court, London, SW7 2QJ, United Kingdom
Printed in Croatia

British Library Cataloguing-in-Publication Data
A catalogue record for this book is available from the British Library

Additional hard and PDF copies can be obtained from orders@intechopen.com

Salmon Aquaculture
Edited by Qian Lu
p. cm.
Print ISBN 978-1-83969-617-6
Online ISBN 978-1-83969-618-3
eBook (PDF) ISBN 978-1-83969-619-0

For EU product safety concerns:
IN TECH d.o.o., Prolaz Marije Krucifikse Kozulić 3, 51000 Rijeka, Croatia, info@intechopen.com or visit our website at intechopen.com.

Meet the editor

Qian Lu received his Ph.D. from the University of Minnesota (USA) and then worked at Nanchang University (China). Qian Lu is doing research on aquaculture, including innovative aquaculture models, nutritional values of aquaculture feed, green technology for aquaculture and the remediation of eutrophic wastewater, and the ecosystem of aquaculture. Some of his research has been successfully applied in aquaculture to address technological problems.

Contents

Preface

Salmon aquaculture is becoming more prevalent due to the high popularity of salmon products in the market. However, the development of salmon aquaculture is faced with a variety of problems. This book provides some new ideas about salmon aquaculture, such as the use of probiotics to prevent antibiotic resistance and enhance the immune response of salmon. In addition, the book discusses some novel models of salmon aquaculture. This volume is a useful contribution to the field of salmon aquaculture.

Qian Lu
Nanchang University,
Nanchang, China

Section 1

Novel Technologies for Salmon Culture

Chapter 1

New Development: High-Strength Stainless Steel as a Sustainable Material for Aquaculture

Paul Gümpel, Urs Dornbierer, Arnulf Hörtnagl and Torsten Bogatzky

Abstract

This paper presents the current state of development and selected technological challenges in the application of ecologically and economically sustainable nets for aquaculture based on ongoing development projects. These aim at the development of a new material system of high-strength stainless steel wires as net material with environmentally compatible antifouling properties for nearshore and offshore aquacultures. Current plastic netting materials will be replaced with high-strength stainless steel to provide a more environmentally friendly system that can withstand more severe mechanical stresses (waves, storms, tides and predators). A new antifouling strategy is expected to solve current challenges, such as ecological damage (e.g., due to pollution from copper-containing antifouling substances or microplastics), high maintenance costs (e.g., cleaning and repairs), and shorter service life. Approaches for the next development steps are presented based on previous experience as well as calculation models based on this experience.

Keywords: sustainability, stainless steel, high strength, redesign, predator net

1. Introduction

Of the fish consumed worldwide, the total amount since 2016 has been more than 150 million tons per year, almost half is now produced in aquacultures. While the amount of freely caught fish is stagnating or slightly declining, the amount of fish farmed in aquacultures is continuously increasing. Currently, about 150 species of fish are farmed in aquacultures, and the proportion of seawater farms is constantly increasing [1].

A crucial challenge of today's aquaculture is to protect the farmed fish but also to protect the environment and the surrounding ecosystems. The goal of sustainable aquaculture operations should therefore be to achieve as limited an impact on the environment as possible. At the same time, in the sense of economic operation, care must be taken to ensure that the fish farms are subject to as limited a maintenance and repair effort as possible. In addition, the conditions under which the fish are kept play a decisive role, so that the fish do not suffer from diseases and a sufficient exchange of water is ensured.

The massive expansion of fish farms poses major challenges for operators. Sustainable criteria for aquacultures are indispensable and it is important to avoid negative impacts on the environment in order to meet consumer confidence as well as current and future national and international requirements. The materials used for the construction of the nets play a major role. Generally, a net enclosure consists of a buoyant support system and a net that encloses the animals. There are only minor differences in the basic principles between the classic "nearshore" and the "offshore" applications that have been implemented for a few years. However, the individual installations can vary greatly in size, shape, and materials used [2, 3].

Plastics are predominantly used as the netting material. To counteract the growth of undesirable organisms on the surface of man-made structures, also better known as biofouling, these plastic nets are usually protected from fouling with anti-fouling agents. These industrially available products are currently often made of copper and zinc. It should be noted that both metals are listed by the EU under the Dangerous Substances Directive (67/548/EEC). Especially the application of copper in the field of anti-fouling has a significant impact on the fish in the fish farm as well as on the organisms in the direct environment of the fish farms [4, 5]. Thus, significantly elevated concentrations of copper in sediments can be detected in the vicinity of salmon aquaculture operations that use conventional anti-fouling strategies with copper. Individual scientific studies also indicate that the high concentrations of copper may also lead to sub-lethal effects for the fish kept in the aquacultures [4–6]. The application of copper-containing solutions to the net materials used is mainly by copper-containing dip coatings, which impregnates the nets. This results in a continuous release of copper into the environment, causing the effects listed above to occur. For example, in the OSPAR Commission reporting area, this resulted in an annual emission of at least 454 metric tons (MT) of zinc, copper, and chemical compounds based on them as early as 2009 [7]. Due to the large increase in aquaculture since 2009, a much higher number can be expected today on a global scale. These considerable amounts of valuable raw materials are consumed in this application and cannot be directly recovered by conventional recycling. The search for suitable and practical alternatives to this end has been ongoing in the field of research for several years [8]. Nevertheless, the use of anti-fouling strategies based on copper and zinc remains the state of the art.

Moreover, the polymer materials used are hardly recyclable and in turn also contribute to plastic pollution of the oceans. Plastics, and in particular the fine fibers of the nets or the ropes and braids, are also very susceptible to the biofouling already mentioned above (substance deposits on the material surfaces). They therefore require complex and costly maintenance and cleaning. In addition, the susceptibility to fouling places a burden on the fish living in the aquacultures. This in turn leads to costly treatment, which also stresses the fish and slows their growth. Omitting the substances that protect against anti-fouling leads to a considerable reduction in the cross-section of the nets within a short time due to the colonization of algae, mussels or barnacles. This fouling can adhere very well to conventional plastic nets and can only be removed with great technical effort, if at all. In addition, the fouling on the nets leads to a significant increase in the mass of the net systems. This results in a significantly higher mechanical load and, in conjunction with the other loads that occur (wave action, tides, currents), can also lead to corresponding damage to the nets [9].

Damaged nets, but also nets that can be easily bitten through by predators, do not provide sufficient protection against classical predators of fish in aquaculture. The often-used setup with two net systems (predator net to keep predators away and fish net inside the aquaculture) is also only limited suitable to keep away the highly specialized and intelligent predators. Exemplary for many other cases are seals as a challenge for salmon farms in Scotland. An adult harbor seal is an extremely efficient hunter, eating between 3 and 7 kg of fish per day, depending on species and size. In addition, continuous attacks by predators in the aquacultures lead to a significantly increased stress level of the fish, which not only inhibits their growth, but can also lead to their death. The sometimes widespread method of selectively hunting and killing corresponding predators in the environment of aquacultures is restricted by increasingly restrictive regulations at national and international level, so that there is a need here for innovative and sustainable methods to protect fish in aquacultures [10, 11].

All aspects listed here suggest that for an ecologically and economically sustainable operation of aquacultures, both "nearshore" and "offshore", it is necessary to resort to further technological approaches and concepts. The application of high-strength metallic mesh systems represents a promising alternative technology in this respect. The state of the art, the development work carried out to date and the prospects are reported on below.

2. Material selection and mesh production

The selection and configuration of the net materials as well as their long-term behavior in seawater determine decisive parameters for aquaculture operations. As already described above, the flow of fresh water through the cages, the hygiene of the facility, the protection against external predators and, last but not least, the safety against escape of the farmed fish into the open sea are all significantly influenced by this. For all the above criteria, stainless steel is an ideal material for several reasons. The high specific strength of steel compared to polymers also makes it possible to produce larger net systems and increases the water flow. The technical challenge of producing a mesh from steel wires has now been solved very well on an industrial scale. It seems to be important that only the pure stainless steel surface is used without any coating or that no toxic substances can be used to prevent fouling of the stainless steel surface. The great advantage of stainless steel is the good and easy removal of biological fouling. To build a complete plant with stainless steel cages is in principle technologically feasible, but still represents a major challenge. In particular, the design of the plant, the connection of the nets and the associated force transmission, as well as the handling of the nets must be adapted to the properties of the stainless steel.

In order to find a fundamentally suitable material for aquaculture nets in nearshore and offshore fish farming, several development steps were carried out. The first step was to compare the properties of different stainless steels to meet both wire and net manufacturing and equipment operation requirements. The properties of different types of stainless steels were investigated and evaluated. The sometimes contradictory requirements for high strength with sufficient residual ductility for the manufacture of nets and, in particular, the corresponding corrosion resistance for use in seawater can best be met by the so-called corrosion-resistant duplex stainless steels (**Figure 1**).

Duplex stainless steels also offer the advantage of high resistance to stress corrosion cracking in seawater. In the oil and gas industry, these steels have been used in seawater for many years with very good experience.

Replace the entirety of this text with the main body of your chapter. The body is where the author explains experiments, presents and interprets data of one's research. Authors are free to decide how the main body will be structured. However, you are required to have at least one heading. Please ensure that either British or American English is used consistently in your chapter.

By using duplex stainless steels in the cold-worked condition, all requirements regarding mechanical strength and corrosion resistance can be met. Laboratory

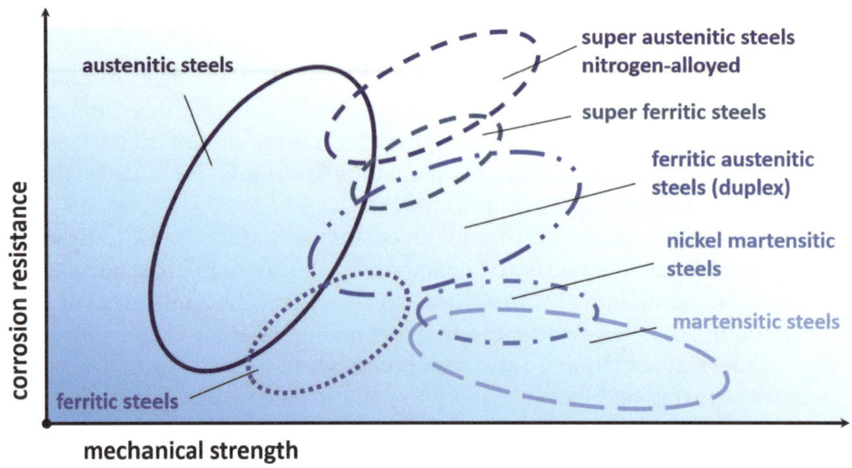

Figure 1.
Schematic overview of the relation between corrosion resistance and mechanical strength for the different types of stainless steels [12].

tests have shown that both materials A (alloy 2304 (UNS S 2304/ 1.4362)) and B (alloy 2205 (UNS S2205/1.4462)) have corrosion resistance in artificial seawater at the usual temperatures. However, the resistance of the molybdenum-containing material B is higher, resulting in greater safety against additional stress factors such as higher temperature, concentration of chloride and/or biologically influenced effects on the corrosion process. A highly specialized manufacturing process was used to produce and further process meshes made of these high-strength materials with a tensile strength of up to 2000 MPa. Therefore, the mesh system manufactured from stainless steels has much higher mechanical strength compared to the conventional plastic meshes. The use of stainless steels with the much higher strength compared to polymers results in thinner mesh bars or larger mesh sizes. The area ratio of web material to flow opening becomes more favorable and water exchange, which is so important for the quality of the plant, becomes much better compared to polymer nets. Regardless of the degree of biological growth, this improves the living conditions of the fish. This positive condition is maintained by a continuous cleaning process, which leads to a higher sustainability.

3. Practical testing

A selection of different net systems (material and antifouling strategy) represented the second step of the development carried out. For this purpose, samples were outsourced at eight locations worldwide (**Figure 2**) over a period of at least 6 months in order to investigate the individual fouling behavior in comparison to existing net systems in practical use. For this purpose, samples were outsourced in the area of fish farms or shellfish farms in order to simulate local conditions in the farm area as well (see **Figure 3**). Contamination was documented and evaluated at defined intervals using photography and light microscopy. In addition, the cleanability of different mesh systems was tested and evaluated using standardized cleaning procedures.

After several test cycles, the samples were evaluated for their corrosion and antifouling behavior. In addition to these immersion tests in natural fish farm environments, laboratory tests as well as microbiological and corrosion tests were performed to investigate and evaluate the different net systems and AF strategies.

Figure 2.
Sites for natural outsourcing in aquaculture and in artificial seawater for simulation experiments in the laboratory.

Figure 3.
Test arrangement of the outsourced samples.

The results of the laboratory tests as well as the natural deposition in seawater farms clearly showed that the development and also the growth of biofilms (**Figure 4**) is most safely hindered by the application of toxic substances, e.g. of copper as a common antifouling strategy for a limited time. In polymer nets, this is achieved by infiltration of copper. Metal nets, for example, may be made of copper alloys, or the surface of the steel may be coated with a layer of copper or copper alloys. Surface coatings of non-stick materials such as PTFE (polytetrafluoroethylene) or nanostructured materials (sharkskin effect) can reduce biofouling compared to the surface of pure steel, but fouling is only delayed but not prevented (**Figure 5**).

Another essential part of the investigations carried out was the comparison of the possibilities to clean the overgrown structures after a defined aging period. Different cleaning systems were used for these tests. The results can be described as follows: Steel/metal surfaces can be cleaned thoroughly and almost residue-free with a water jet/hydrojetting (see **Figure 6**). Cleaning with this common method is also possible in principle for polymer meshes. However, a closer examination of the surfaces showed that the cleaning result is significantly worse. In this context, biological studies have shown that considerable residues of biofilm material always remain between the individual plastic threads. Such residues of biological material accelerate re-biofouling. Thus, in these cases, a comparatively rapid re-growth of the nets is to be expected. This effect sometimes leads to a reduction in the service life of conventional plastic nets. A comparable acceleration of re-biofouling could not be determined for the steel nets used.

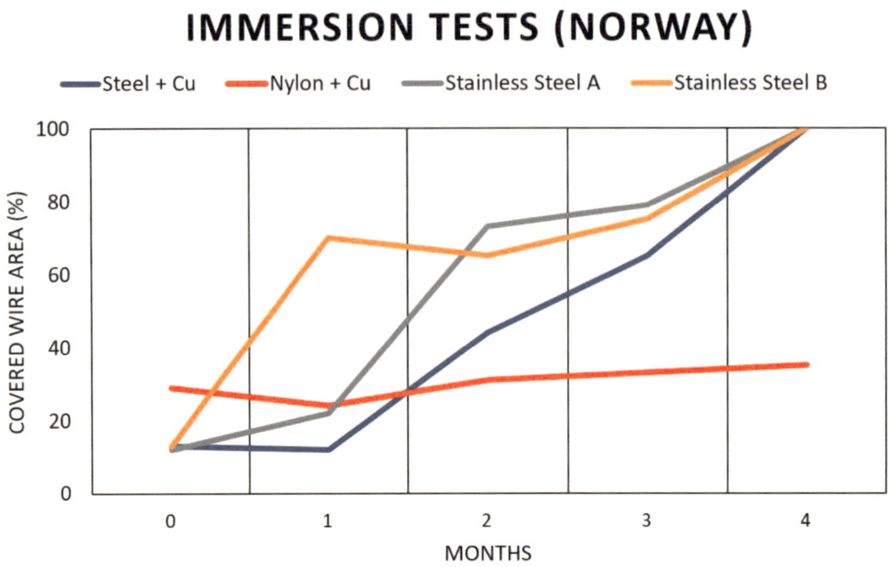

Figure 4.
Proportion of the covered area after long-term exposition in seawater (4 months).

Figure 5.
Method for optical evaluation of the change in the area fraction of the biofilm on the outsourced network structures [12].

On the basis of these preliminary investigations and the material selection made, the first stainless steel nets for aquaculture could be used as predator nets. These coarse-meshed nets keep predators such as seals away from the fish nets inside. The first field trial took place in South America on the Pacific coast. **Figure 7** shows the supplied net rolls that were assembled into cages in the field. Based on the results of the previous laboratory and exposure tests, the nets were made from Material A.

Surprisingly, after only a few months of real-world application in the Pacific Ocean, corrosion attack occurred mainly at the nodes of the mesh (**Figure 8**). Some isolated corrosion attacks were caused by small, invisible defects in the surface. In general, the surface quality of the cold-drawn wires in this highly reinforced condition is a problem. These manufacturing problems were solved by the steel supplier.

The reason for the systematic failures in the node areas could be determined by fault analysis and simulation of corrosion experiments. During field operation, there is high friction at the nodes in the meshes. This friction in the nodes under

Figure 6.
Proven cleaning method to remove biofouling from surfaces.

Figure 7.
Installations in South America - Total 112'000 sqm.

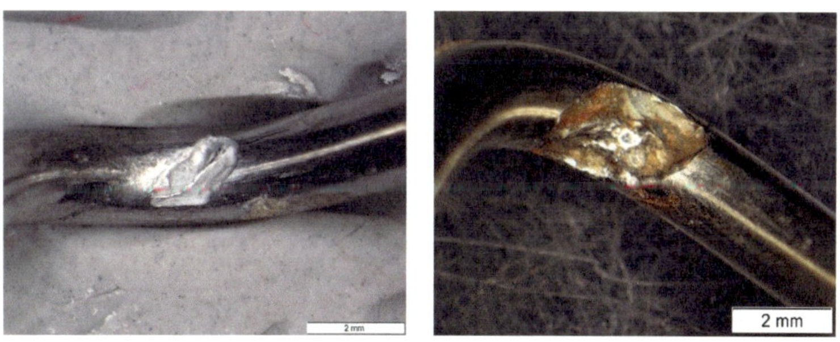

Figure 8.
Typical tribocorrosion in the friction area of a stainless steel net after some months in the Pacific Ocean [12].

high mechanical stress and seawater environment has led to tribocorrosion. In this particular case, the material in the friction area was activated and could not passivate again under the given conditions. Simulation tests showed the activation. The OCP breaks down during frictional loading and increases when friction stops

(**Figure 9**). A dependence on the time between activation is also evident. The tests showed very clearly that material B has a much higher resistance than material A under these test conditions (**Figure 10**). Temporary activation can occur with material B, but repassivation occurs quickly and reliably. The stainless steel with the higher alloy content and thus a higher pitting resistance value definitely has a higher tolerance to any type of corrosion [12].

The susceptibility of individual materials to the tribocorrosion that occurs led to a further and more detailed investigation of possible locations for aquacultures. Using specially developed measuring buoys, it was possible to compare the corrosivity at different locations. This clearly showed that, in addition to influencing factors such as water temperature or salt concentration, the potential shift caused by microorganisms, also better known as ennoblement, plays a decisive role in the corrosion mechanisms that occur. Since, as already explained above, a correspondingly high input of microorganisms is always to be expected in the direct environment of aquacultures, the ennoblements must always be taken into account when selecting materials. As already listed above, this is possible with the listed alloy B, which could be confirmed by laboratory tests as well as by application in aquacultures [12].

In order to better understand the advantages but also the application limits of steel nets compared to plastic nets, a consideration of specific technical properties is necessary. However, it is important that properties such as density are not considered in isolation. The density of a conventional plastic such as PA 6 (polyamide 6) is 1.1 g/cm^3, whereas the density of a stainless steel is approx. 7.8 g/cm^3. This serious difference in density is put into perspective when the achievable tensile strength of these materials is taken into account. For reconditioned PA 6 ropes, a considerable tensile strength of 110–175 N/mm^2 can be achieved. A work-hardened duplex

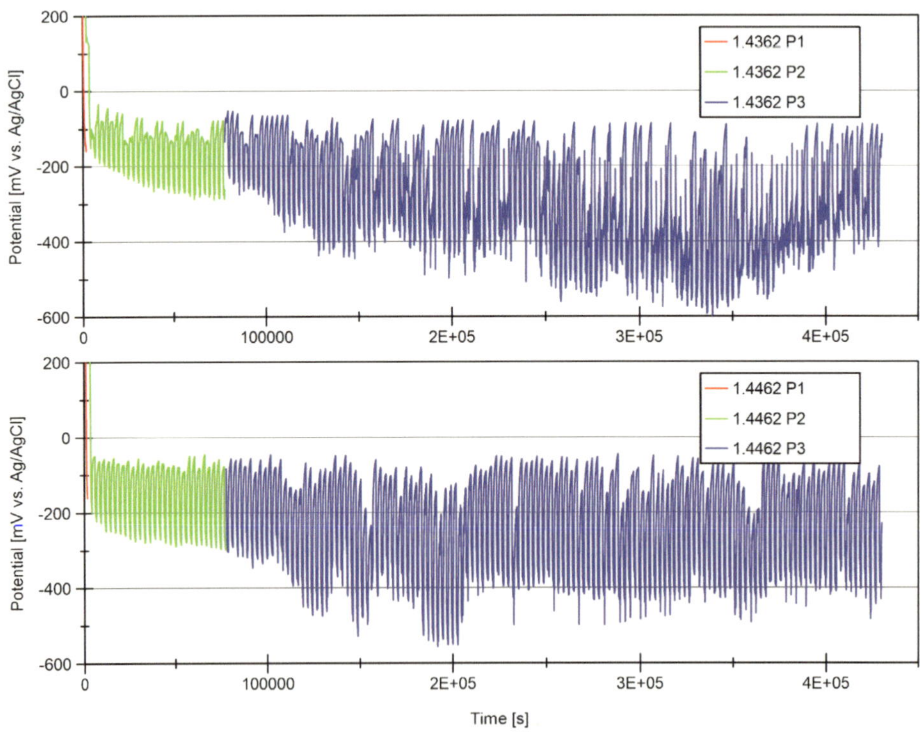

Figure 9.
Open circuit potential during friction simulation – peaks show repassivation when friction stops [12].

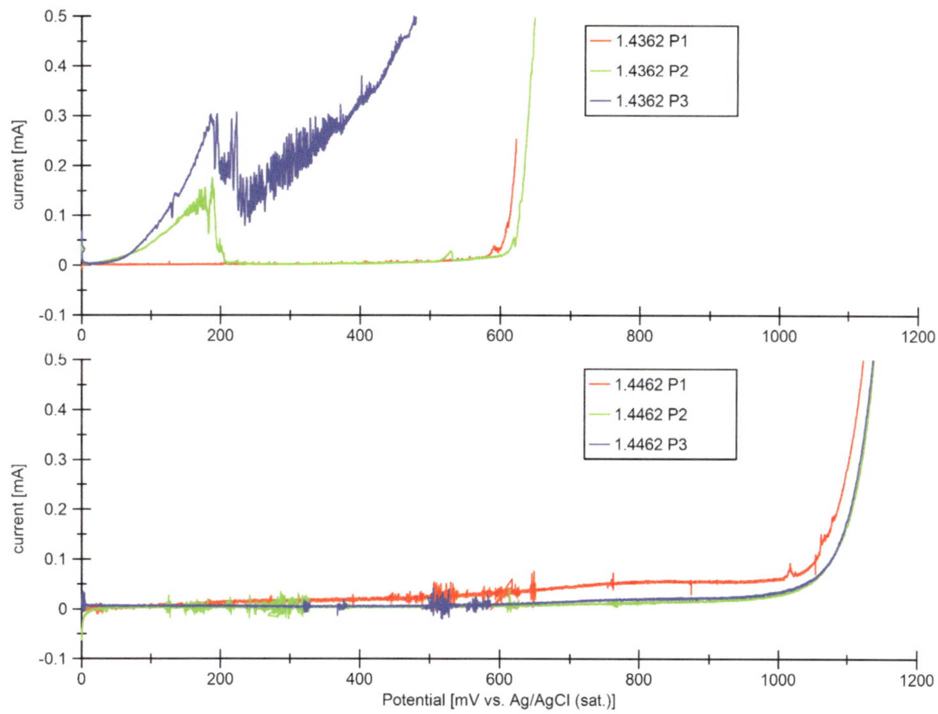

Figure 10.
Current-potential-curves after different times of friction simulation (see phases in ***Figure 9****) [12].*

stainless steel 1.4462 achieves tensile strengths of 1200–1350 N/mm^2, in the case of the highly specialized applications listed here up to 2000 N/mm^2. If these figures are put in relation to each other, the strength achieved per mass used is comparable for both steel and plastic.

Another significant difference lies in the modulus of elasticity of the materials. For the PA 6 material mentioned above as an example, this is approx. 3 GPa. For the steel 1.4462, the value is approx. 200 GPa. This difference is of particular importance, as it clearly shows that a simple substitution of plastic meshes by high-strength steel meshes can be problematic when using the same joining technology. **Figure 11** shows an example of the deformation occurring in a salmon aquaculture. It can be clearly seen that in the application, a significant deformation compared to the original prefabricated geometry occurs due to the dead weight of the mesh panels.

Subsequently, the occurring sea loads, which are composed of the external influencing factors (e.g. water current, wind and tides), result in a time dependency of the occurring forces and thus also of the occurring deformation. This can be seen from the differences in the occurring deflection recorded by measurement on three consecutive days in **Figure 12**.

While conventional plastic nets achieve high deformation even at low forces due to the comparatively low modulus of elasticity of the materials used, steel nets behave comparatively rigidly in most cases. As a result, the load distribution of plastic nets is generally more benign and stress peaks are evenly reduced, even when the mesh panels are rigidly connected to each other. If a correspondingly rigid connection is made in the case of steel nets, there is a risk of wire breaks occurring in the area of load application or at connection points between the nets. This can be the case both as a result of a one-time overload and as a result of a high cyclic load. Taking into account the corrosive environmental conditions in

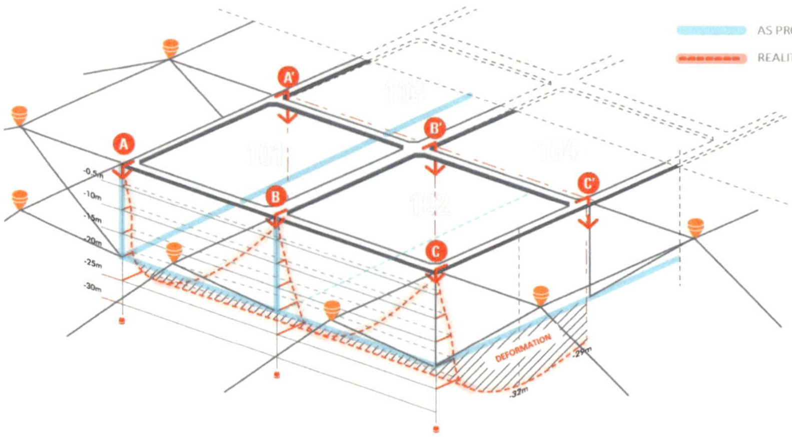

Figure 11.
Illustration of the occurring deformation of an exemplary salmon aquaculture.

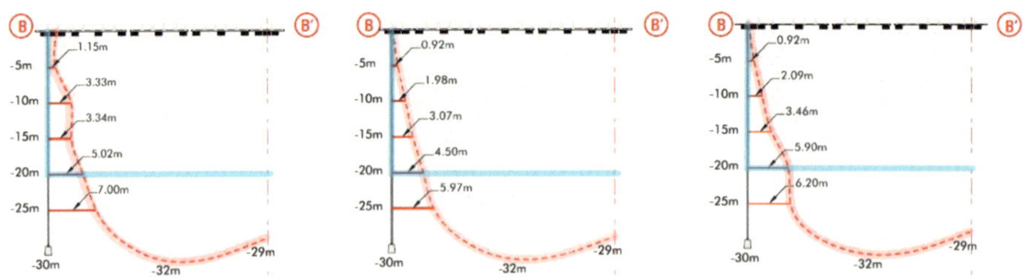

Figure 12.
Temporal variation of the occurring deformation in an aquaculture.

seawater and the tribological stress occurring at the contact points between the individual wires, the result is a highly complex system in which several damage mechanisms interact.

Before remedial measures can be defined, it is necessary to differentiate the damage patterns and the causes of damage as far as possible. As an example, the damages in **Figures 8** and **13** can be compared. Basically, both damage patterns show a reduction in cross-section in the area of the wire contact. However, tribocorrosion in the contact point, see **Figure 8**, leads to increased local corrosion of the surface, while mechanical overloading leaves a comparatively smooth contact point. The wear marks shown in **Figure 13** do not reveal any discernible corrosive attack. It should be noted that the wire breakage that occurs as a result of the load distribution is not directly present at the generated groove. Due to the geometry of the half mesh or the meshes used, the maximum stress occurs directly next to it.

More in-depth investigations on a laboratory scale and simulations based on these confirm that a simple substitution of plastic nets by steel nets in aquaculture with strong mechanical loads due to waves, wind and tides does not yet produce the fully desired increase in the targeted service life when using steel nets compared to plastic nets. As shown by the study of a wire break **Figure 14** of a net element tested for continuous load, violent bridging in the form of sliding or honeycomb fractures occurs as a result of the bending load.

However, the studies carried out and the first applications in practice also show that there is considerable potential in the technology described here. In order to do justice to this potential in the growth market of aquaculture, selected development approaches are listed in the following chapter.

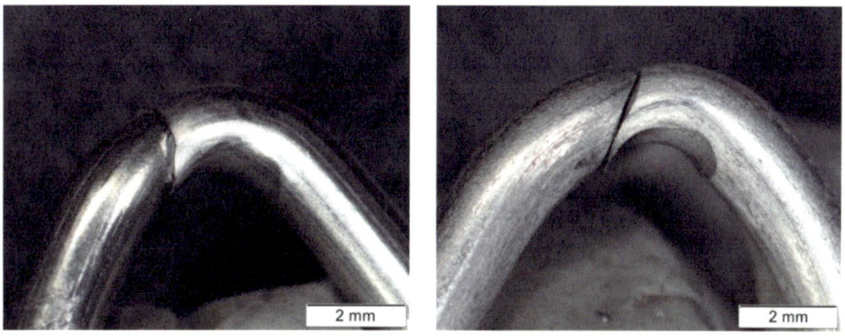

Figure 13.
Fracture pattern at a half mesh in the area of the wire contact of two different mesh patterns due to mechanical overload.

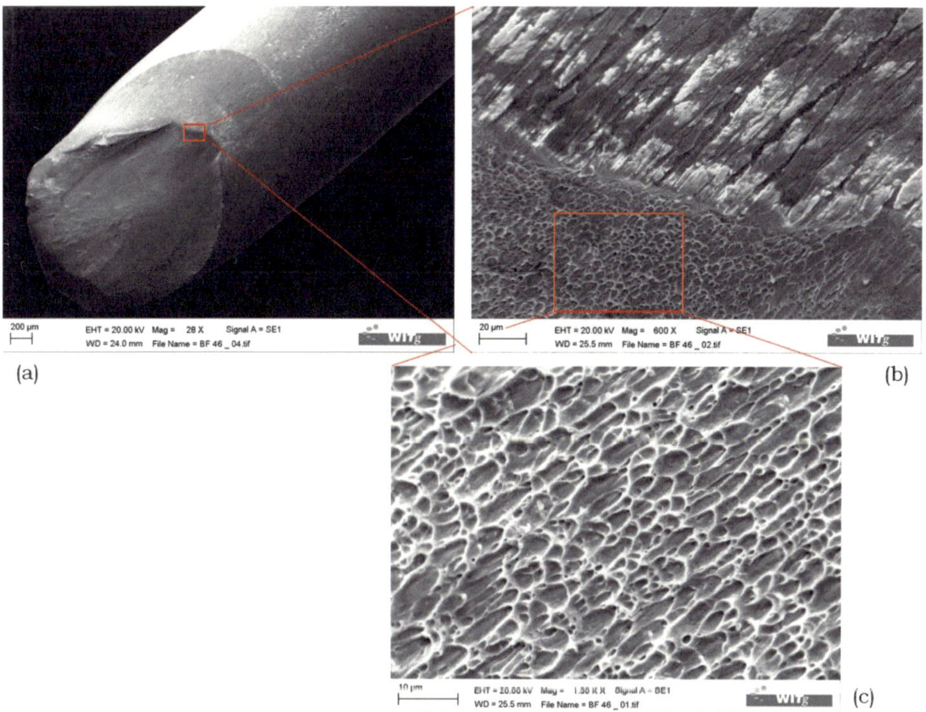

Figure 14.
SEM images of the fracture surface of a wire at different magnifications; (a) overview image showing presumed area of origin for the fracture; (b) and (c) detail images showing characteristic honeycombs of a transcrystalline honeycomb fracture.

In order to round off the comparison between plastic nets and steel nets, the input of plastic/microplastic into the waters must be taken into account in addition to the influence of the antifouling strategy applied. Similar to steel nets, the mechanical stresses that occur between the individual ropes and strands lead to wear. This wear occurs when plastic nets are used for both conventional fishing and aquaculture. It must be pointed out, however, that this input into the world's oceans is comparatively small compared with other sources. Nevertheless, from a current perspective, it is particularly important to reduce the generation of microplastics in the direct vicinity of our food sources [13]. While plastic nets must be disposed of as hazardous waste at the end of their service life due to contamination with heavy metals and biomass, steel nets can be recycled very easily as high-quality steel scrap.

4. Outlook and ongoing developments

Based on the above experience, selected changes have been made which should make a significant contribution to the possible widespread use of steel nets in aquaculture. The most significant change in the manufacture of nets is the use of the more corrosion-resistant material B, as already explained above. In addition, the structure and overall suspension of the cage were designed to minimize friction in the nodes and to have the most homogeneous load distribution possible within the net. After these modifications, no systematic signs of corrosion or tribocorrosion were observed. Another starting point is the structural adaptation of the fastening of the nets, the load application and the connection of the nets to each other. A project underway for this purpose is now in a phase in which long-term operation is being observed and a continuous improvement process is being carried out in actual use. A long-term tested, technically reliable solution for anti-predator networks should be available by mid-2022.

In parallel to the above mentioned tests, a development project for a sizing tool for farms to be equipped with stainless steel netting is ongoing. Since local influences differ greatly depending on the location of the aquacultures (prevailing currents, wave action, tidal variation and the factors listed above that affect corrosion), it is necessary to evaluate each aquaculture individually. Depending on the stress factors encountered, other design measures can then be taken. The tool will allow statements on the design of the network and its expected service life based on local conditions. This project should be completed by 2025.

Initial tests and simulation on this indicate that a successful use of high-strength stainless steels as network material for aquaculture is possible. However, due to the damage mechanisms that occur, holistic approaches must be pursued, which are aimed in particular at a possible reduction of the acting loads and their uniform distribution. The following approaches can be considered as examples.

By gradually reducing the wire diameter used in steel nets, the weight force occurring in aquaculture can be reduced as much as possible. Calculations show that the use of thinner wire diameters in the lower part of the aquaculture results in a reduction of up to 30% of the mechanical loads in the area of the upper attachment points in steel nets, without any restriction of the function.

In the case of plastic nets, whose density is only slightly higher than that of seawater even when impregnated, the use of weights and attachments to the seabed is necessary to minimize their movement due to currents and tides. These measures can and must be dispensed with wherever possible in the case of steel nets. In this way, the mechanical stress on the aquaculture can be further reduced. At the same time, the higher density of steel nets ensures less deformation of the cages.

The use of flexible connecting elements between the mesh panels allows a more even load distribution, which reduces stress peaks. These stress peaks, which can lead to fatigue failure especially with a high number of load cycles, are thus defused as critical points and a much more homogeneous load distribution is achieved. As can be seen from the example in **Figure 15**, constructive approaches can be used in the design of the network paths. Although this requires a corresponding know-how in the manufacturing and processing technology of rope systems, the simulations carried out as well as the first field tests show a significant reduction of the occurring stress peaks.

Another measure is the application of a hybrid structure of aquaculture. As already mentioned above, it must always be taken into account when selecting materials that materials from different groups in particular differ in a variety of

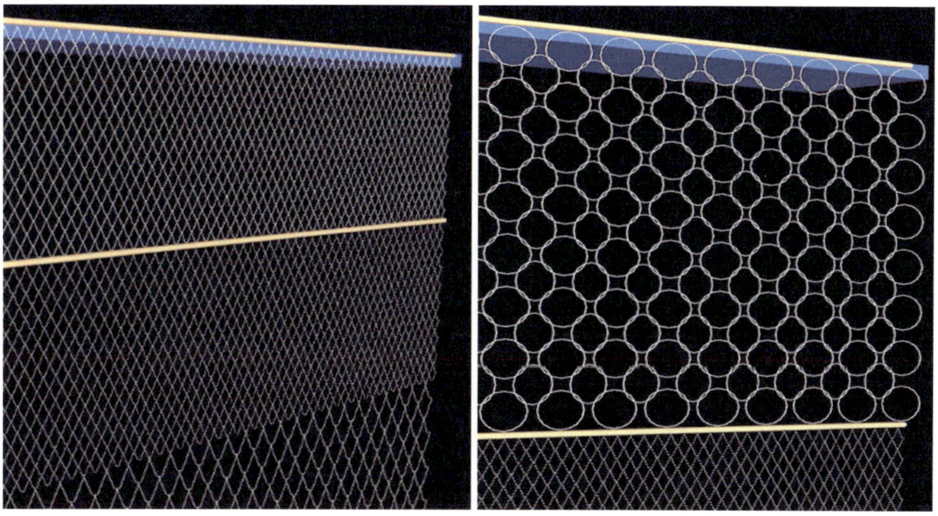

Figure 15.
Exemplary presentation of constructive design possibilities in the application of steel nets.

Figure 16.
Installed predator nets for salmon aquaculture in Chile.

properties. A hybrid structure provides for the use of elements made of plastic and elements made of steel. The elements made of plastic, which are used to transfer the load between the floats and the steel nets, allow the load to be distributed as evenly as possible. If the bottom of the aquaculture is also designed as a plastic net, it remains correspondingly flexible or easily deformable. If the side elements are made of steel nets, they continue to offer appropriate protection against predators and are easy to clean and do not require any further AF coating (**Figure 16**).

5. Conclusion

The use of nets made of high-strength and corrosion-resistant steel offers corresponding advantages over conventional nets made of plastic, particularly from an ecological point of view, and subsequently also from an economic point of view after a sufficient service life. The scientific studies carried out show that the challenges of corrosion and tribocorrosion in seawater, biofouling and the mechanical stresses that occur can be met by selecting a suitable stainless steel material. However, it should be remembered that the exclusive use of steel nets for large-scale conventional aquaculture is not possible at present. This challenge can be met in the future with the approaches taken for further development and adjustments in the connection and force application of the nets, according to the available data of these challenges.

Regarding the common aquacultures in the near shore area (near shore), two trends are clearly visible in the past years: off shore and on shore. It is self-evident that for most on shore aquacultures do not require a corresponding network infrastructure. For offshore aquacultures, on the other hand, new challenges and opportunities arise through the application of the high-strength steel nets presented here. The comparatively high tensile strengths of the materials used, up to 2000 MPa, make it possible to operate aquacultures safely even under comparatively harsh conditions if the overall design is adapted.

Thanks

Special thanks to the Swiss Agency for the Promotion of Innovation - Innosuisse for the financial support of the research work carried out.

Author details

Paul Gümpel[1], Urs Dornbierer[2], Arnulf Hörtnagl[3]* and Torsten Bogatzky[3]

1 Konstanz University of Applied Sciences, Konstanz, Germany

2 Geobrugg AG, Romanshorn, Switzerland

3 Institute for Materials Systems Technologies Thurgau, Tägerwilen, Switzerland

*Address all correspondence to: a.hoertnagl@witg.ch

References

[1] Food Outlook – Biannual Report on Global Food Markets. FAO 2020.

[2] Wang CM. Moving offshore for fish farming 2019.

[3] Holmer M. Environmental issues of fish farming in offshore waters: perspectives, concerns and research needs. Aquacult. Environ. Interact. 2010; 1(1): 57-70. DOI: 10.3354/aei00007

[4] Nikolaou M, Neofitou N, Skordas K, Castritsi-Catharios I, Tziantziou L. Fish farming and anti-fouling paints: a potential source of Cu and Zn in farmed fish. Aquaculture Environment Interactions 2014; 5(2): 163-71. DOI: 10.3354/aei00101

[5] Børufsen Solberg C, Sæthre L, Julshamn K. The effect of copper-treated net pens on farmed salmon (*Salmo salar*) and other marine organisms and sediments. Marine Pollution Bulletin 2002; 45(1-12): 126-32. DOI: 10.1016/S0025-326X(01)00296-X

[6] Dean RJ, Shimmield TM, Black KD. Copper, zinc and cadmium in marine cage fish farm sediments: an extensive survey. Environmental Pollution 2007; 145(1): 84-95. DOI: 10.1016/j.envpol.2006.03.050

[7] OSPAR COMMISION, editor. Assessment of Impacts of Mariculture: Biodiversity Series; 2009.

[8] Hanno Schnars CB. Biozid-freier Bewuchsschutze für die Marikultur: no copper. Projektendbericht zum Projekt az29456, Institut für Marine Ressourcen Mai 2014.

[9] M. Baum. Umweltfreundliche Netze für die Fischzucht. Kiel: Forschungsschwerpunkt Kiel Nano Surface and Interface Sciences; 2020 Jul 21.

[10] Global Aquaculture Alliance. Fish farmers seek non-lethal deterrence for persistent predators « Global Aquaculture Advocate; 2021 [cited 2021 May 26] Available from: URL: https://www.aquaculturealliance.org/advocate/fish-farmers-seek-non-lethal-deterrence-persistent-predators/.

[11] Seals and Tasmanian Aquaculture. Tasmanian Conservation Trust 2015 Jul 10.

[12] Gümpel P, Hörtnagl A, Sorg M. High Tensile Stainless Steel as a Sustainable Material for Aquaculture. Procedia Manufacturing 2019; 30: 315-22. DOI: 10.1016/j.promfg.2019.02.045

[13] Microplastics in fisheries and aquaculture: status of knowledge on their occurrence and implications for aquatic organisms and food safety 2017.

Chapter 2

Integrated Culture of *Oncorhynchus mykiss* (Rainbow Trout) in Pre-Cordilleran Sector under a Recirculation System in Northern Chile

Renzo Pepe-Victoriano, Héctor Aravena-Ambrosetti and Piera Pepe-Vargas

Abstract

An experience of integral farming of *Oncorhynchus mykiss* (rainbow trout) is carried out in Copaquilla, 90 kilometers inland from the city of Arica at 3,000 mamsl. The system used was the Recirculating Aquaculture System (RAS), which had six ponds of 40 mt^3 each, two decanters with a capacity of 3.5 mt^3 and a biofilter of 3.5 mt^3 with substrate for the fixation of ammonium and nitrite transforming bacteria. The three latter ponds were buried below the lowest level of the fattening ponds. Three pumps, two running and one 1.5 hp. backup, plus a 1 hp. blower, were the water and air equipment utilized in the system. Each pump had a flow capacity of 450 lt min^{-1}. This water was sucked from the biofilter and transferred to the accumulator tank with a capacity of 10 mt^3. From there it was distributed by gravity to the fattening ponds. In addition, the juvenile system had a particular SAR with a 0.5 hp. pump, a small 0.2 hp. blower and an 80 watt UV lamp. The grow-out SAR received 6,000 trout with an average weight of 15 grams. The group reached approximately 1,200 grams over a year. Thirty fish were selected for reproduction. Eggs were obtained, followed by fry, juveniles and adults. This initiative demonstrated the effectiveness of producing trout in the foothills of the interior city of Arica, Chile.

Keywords: indigenous communities, water quality, rainbow trout transport, spawning, rainbow trout eggs and larvae

1. Introduction

The lack of opportunities, technological improvements and diversification are issues to be resolved in mountain and foothill sectors and the reason for the insertion of new productive and sustainable alternatives. The unbeatable environmental conditions that many of the sectors of this territory present, such as the existence of good quality water sources, availability of space and microclimatic environments, are clear elements of potential. In such conditions, aquaculture shows favorable prospects for development.

Aquaculture, as indicated by Pepe-Victoriano et al. [1], is presented as a real productive alternative, by identifying crops of freshwater species of commercial value as a way to take advantage of the capacities installed in the agricultural activity of the area. Storage ponds for irrigation, greenhouses, hydraulic systems, among others, are some examples of facilities that add to the possibility of reusing water as many times as necessary before being derived as a final destination for irrigation of plants and vegetables.

The utilization of technology such as recirculation in aquaculture systems is presented as a powerful alternative for pre-cordillera areas [2]. A recirculation system allows more precise control of the main environmental parameters [3]. Water temperature, to name one, is a critical parameter for most poikilothermic organisms, such as fish, and can be controlled much more economically in a recirculating system than in an open flow one. Controlling this and other environmental parameters allows for faster growth and more efficient use of feed due to reduced stress on the farmed organisms.

The quality of the water in Copaquilla, as the main variable for trout farming, is within the optimal range for the development of this initiative. In previous studies to the three existing springs that would feed the farm, elements such as arsenic, copper, zinc, iron, and manganese are within the limits allowed by Chilean standards for drinking water consumption and therefore for freshwater crops.

Moreover, it is worth mentioning that trout production in recirculation systems would make it possible to increase the supply of good quality fish for local and regional population. Per capita fish consumption in Chile averages 13 kilos per year, which keeps us quite far from the world average of 20 kilos established by the Food and Agriculture Organization of the United Nations [4]. It is estimated that the unmet demand for fresh fish for human consumption in our region is around 1,000 tons per year. In addition to this, there is a growing consumption with clear preferences for salmonid fish species over red meat and poultry, mainly in the young segment of the population.

The implementation of this fish farm generates clear potential and economic benefits for the communities in the foothills of the Andes that might develop such initiative further into the future. This development is mainly projected as a diversification alternative, adding value to the aquifer resource, and improving the income of the territory by including other economic activities such as tourism and gastronomy, among others. The rich landscape and heritage of the area would also be benefited. In addition, these farming provides useful information to be used as a basis for other farming initiatives in the region. Aquaculture is expected to be boosted and to become one of the main strategic axes of regional development in the short term.

The main objective of this initiative was to develop the integral culture of *Oncorhynchus mykiss* under recirculation conditions, as a productive development alternative for communities in the pre-cordillera of the Arica and Parinacota Region.

2. Why grow trout at 3,000 meters above mean sea level?

The productive development of the communities located in the foothills of the Arica and Parinacota Region, especially in the Copaquilla sector (**Figure 1**), is based mainly on agriculture, livestock and to a lesser extent, tourism. Most recently, activities such as agriculture and livestock have suffered a clear deterioration and abandonment, mainly due to the lack of diversifying alternatives that encourage and avoid one of the main problems of the territory; the migration of young population to the city.

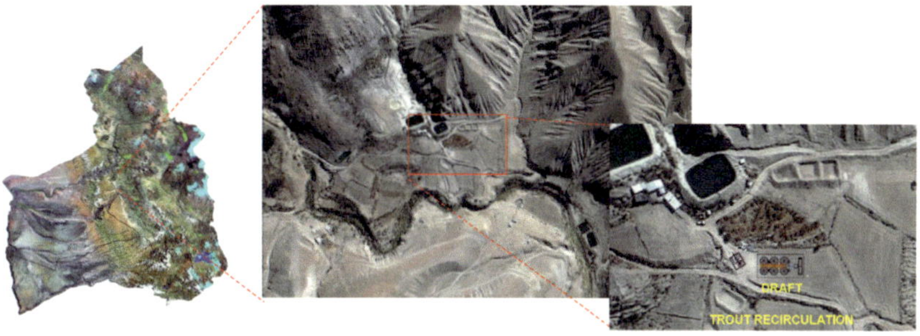

Figure 1.
Geographical location of the Pukara de Copaquilla Cultivation Center (CCPC).

3. Biological characteristics of trout

The rainbow trout (**Figure 2**) is a teleost fish belonging to the salmonid family (Salmonidae), whose distribution range covers cold waters of North America, Asia and Europe. It tolerates temperatures going from 0 to 25° C, with an optimum range of 10 to 14° C to remain healthy. In order to ensure excellent growth, however, temperatures between 15 and 20° C are preferred [6] in good water quality conditions.

The life cycle of rainbow trout is highly variable in terms of migratory patterns. Out in the wild, they spawn in rivers or streams, and many complete their life cycle in freshwater. Some varieties, nevertheless, migrate and spend their adult life in the ocean. They only return to the river where they were born to spawn, completing the cycle. This behavior is known as anadromous reproduction [7]. Anadromous forms migrate to the sea as juveniles and can travel long distances in the ocean. Freshwater forms (non-anadromous) move between affluent and main river, between river and lake, or spend their entire lives in a particular stream or river [7]. The growth and sexual maturation of these organisms can occur in freshwater or seawater. This last phase can last between 1.5 and 3 years of the fish's life. Spawning generally occurs in water flows of both rivers or affluents of the main channel, on gravel beds of rivers or lake shores, where water seeps through it. The gravel provides protection for the eggs until they emerge as fry ready to eat and migrate [7].

Rainbow trout are highly adaptable to their environment, which is why they have achieved a wide distribution [8]. According to studies carried out by other authors, trout born from the same progeny can adapt to totally different habitats. Some can grow, reproduce and live in a small stream, with a few centimeters of water above their bodies. Meanwhile, others can travel many kilometers to the ocean to feed and grow much larger than the first [9]. This is why resident trout, migratory trout from rivers and lakes, or anadromous trout can be found in the same watercourse [9].

Figure 2.
Adult specimen of Oncorhynchus mykiss (Rainbow trout). Image extracted from Cornejo-Ponce et al. [5].

4. Justification for the usage of a recirculation system

Recirculating Aquaculture Systems (RAS) are one of the emerging production technologies that are being applied in the national and global aquaculture sector. This is especially true when considering the concept of efficient improvement in production by minimizing the usage of water resources and increasing environmental responsibility. Thus, concentrating and treating the waste generated during the production process.

This technology is ideal for utilization in production systems that involve fish farming and the implementation of wastewater for agricultural irrigation, since the latter improves productivity and profitability in economic terms. In Chile, this technology is mainly used in the early stages of the salmon production process and is associated with fish farming centers where the incubation phase is carried out until smoltification. In the northern zone of Chile, there are no experiences or farms in operation that involve the application of recirculation technology in production processes with trout or other freshwater species.

The implementation of this type of technology (SAR) in aquaculture has enormous potential, especially in desert areas such as the Arica and Parinacota Region, where water resources are scarce. It has been demonstrated that 90 to 98 percent of water can be reused, compared to traditional open flow farming systems. It also has other advantages, such as: energy savings, maximization of production under water and space limitations, minimization of effluent problems by reducing waste discharges to the environment and controlling and regulating water quality parameters of the crop, among others. On the one hand, experiences in farming demonstrate that these systems can intensively produce up to 25–30 kg/m^3 of rainbow trout and about 80 kg/m^3 of salmon. On the other hand, in semi-closed aquaculture systems with water recirculation, rainbow trout productions of 6,257 kg per year (120 kg per week) have been reported, with a maximum biomass of 66 to 74.6 kg/m^3.

In this innovative technology, some advantages stand out: a) flexibility in the selection of the farming site related to the possibility of using a small amount of water; retaining waste; manipulating the farm medium (especially temperature) and avoiding the entry of organisms from the natural environment or their exit to it, b) biosafety, since it avoids the entry and exit of pathogenic organisms or the exit of specimens of the cultivated species to the natural environment. Moreover, by having direct control over the growing conditions, optimal growing conditions can be kept independent of environmental variations, thus reducing the risk of diseases and ensuring productive yields, c) expedite treatment of certain diseases as it is simpler to handle and minimize the amount of product needed for treatments by immersion, in conditions of relatively high farm density, d) reduction in water consumption, e) reduction in the amount of waste and the possibility of treating it, thus avoiding possible impacts on the environment, f) scaling and replicability in other sectors of the foothills of the region and the rest of the northern side of the country, and g) possibility of integrating renewable energy systems. As of today, there are no viable renewable energy systems due to the volumes and consumption required, but which are perfectly compatible, such as solar energy.

5. Proposed recirculation system for trout farming in the pre-cordillera sector

The closed water recirculation system installed at the CCPC (**Figure 3**) consists of 6 circular Australian-type ponds for intensive trout production. It is provided with a central drainage system and hydraulic connections for water, air and oxygen

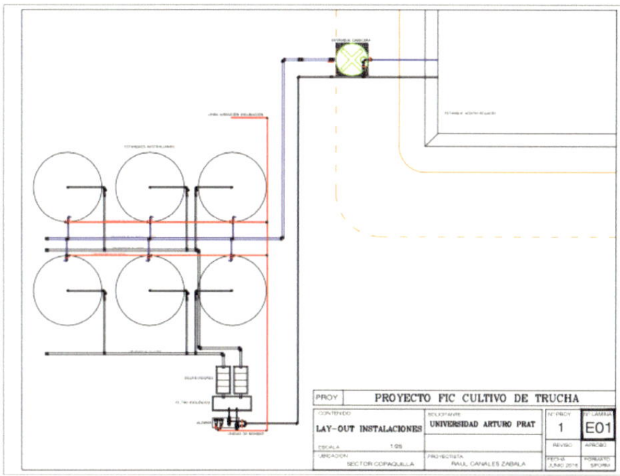

Figure 3.
Schematic of the recirculation system employed in the Rainbow trout farming at the CCPC.

supply; a system of fiberglass ponds located below ground level, including two sedimentation ponds and a pond with biofilters. In addition, two 1.5 hp. water suction pumps, two 1 hp. high pressure blower and an emergency oxygen generator.

The ponds (Australian type) for intensive trout production are made of corrugated galvanized steel with a diameter of 5.4 m and a nominal height of 1.76 m, with a maximum water volume of 40 m^3. The water reaches each pond through a central distribution pipe and lateral outlets, which supply water to each farming unit. The flow is controlled by PVC valves, which, due to their arrangement, generate a circular water movement.

Each pond is internally covered with a 1.0 mm thick black non-toxic liner or geomembrane, which acts as a waterproofing layer. In the center of each pond there is a drain with a 110 mm diameter outlet and a drainage system with 5 mm diameter openings.

In addition, there is an air distribution pipeline that, by means of a blower, allows continuous aeration and oxygenation of the water column to ensure high levels of dissolved oxygen.

Furthermore, there is an oxygen pipeline distributed to all the ponds and that, by means of an oxygen generator will supply a high concentration of this gas, in case of emergency.

The water leaving the six ponds passes through a PVC pipe below ground level until it reaches the sedimentation ponds. They feature an internal division to facilitate the sedimentation of suspended solids.

Waste accumulated at the bottom of the sedimentation ponds is removed periodically through a submersible pump. There, fresh water is added daily to maintain the volume of the pond, which represents approximately 1 to 2% of the total volume of the system. In addition, it has the purpose of removing the final product of the nitrification process (nitrate). Thus, it maintains the water level and compensates for evaporation and handling losses.

Water coming from settling ponds passes into a fluidized submerged biofilter, which contains an ample spectrum plastic substrate for the attachment of nitrifying bacteria. Such bacteria convert the ammonia nitrogen discarded by the fish to nitrites and subsequently to nitrates; molecules that are less harmful to aquatic organisms and, conversely, the main nutrient for most plants.

The nitrification process requires the addition of dissolved oxygen. Reason why this system takes advantage of the volume of air generated by the blower. The diffusion hoses are arranged in such a way that they allow a uniform movement of the biofiltration substrates. By doing so, ammonia nitrogen and oxygen are distributed over the entire surface of the biofilter avoiding the accumulation of solids in the biofilter.

The water from this pond will finally pass to the water accumulator pond, which has a volume of 10 m^3. In this pond the water is aerated before being sent by gravity to the six production ponds.

All power for the recirculation system (two 1.5 hp. pumps, one 1 hp. blower, one 0.5 hp. pump for the juvenile system) was provided by a 7 kw/hour photovoltaic plant, which had 28 photovoltaic panels of 250 watts each, two inverters and 24 batteries. The batteries operated 18 to 20 hours per day, occupying the oxygen generator the rest of the hours.

6. Water quality

The chemical compounds dissolved in the water, as well as other physical factors that affect the water, merge together to form what is known as "water quality". In aquaculture systems, changes in water characteristics that improve the production of a crop should be considered as improvements in water quality,

Parameters	Prior to fish entry into farming ponds	After fish are introduced into the farming ponds	Unit	Methodology
Ammonia	< 0,1	0,12–0,48	mg/L	SMWW 4500-NH3F
Alkalinity	35–43	36–38	mg CaCO3/L	SMWW 2320B
Arsenic	0,24–0,29	0,07–0,16	mg/L	SMWW 3114C
Chloride	65–110	10–70	mg/L	SMWW 4500Cl-
True color	6–8	11–28	Pt-Co	SMWW 2120C
Electrical conductivity	572–631	412–513	μS/cm	SMWW 2510B
Hardness	148–185	101–181	mg CaCO3/L	SMWW 2340C
Nitrate (N-NO3-)	1,63–4,79	21–224	mg/L	SMWW 4500-N B
Organic Nitrogen (N-org)	< 0,01	2–42	mg/L	SMWW 4500-NorgB
Nitrite	< 0,1	< 0,1	mg/L	SMWW 4500-NO_2^-B
Dissolved Oxygen	6,02–7,48	5,78–6,32	mg/L	SMWW 4500-O G
P-H2PO4-	7,25–8,99	21,20–43,19	mg/L	SMWW 4500P-C
pH	6,5–6,6	6,9–8,2	—	SMWW 4500B-H + B
Potassium	7,50–9,00	7,10–8,21	mg/L	SMWW 3111B
Total Dissolved Solids	476–515	257–409	mg/L	SMWW 2540B
Temperature	7–18	7–20	◦C	SMWW 2550B
Salinity	0,28–0,30	0,20–0,25	PSU	SMWW 2520B

Table 1.
Physical and chemical parameters of farming water, before and after trout were introduced.

while those changes that reduce production are a consequence of a degradation of said water quality.

This is given by the combination of physical and chemical properties and their interaction with living beings. With respect to the farming of aquatic organisms, any water characteristic that affects in one way or another the behavior, reproduction, growth, yields per unit area, primary productivity and management of aquatic species is a water quality variable.

Since one of the main objectives of aquaculture is to obtain the best yields possible, it is necessary to be thoroughly aware of ecological conditions in the ponds and the processes carried out there.

Within pisciculture parameters, water quality is of upmost importance. It must have adequate characteristics in terms of quantity (flow) and quality (physical, chemical and biological factors). Physical properties, such as temperature, pH, oxygen, transparency, turbidity, among others, may be subject to sudden variations due to the influence of external factors —mainly atmospheric and climatic changes. Chemical properties, however, are much more stable and their variations are minimal. Only in exceptional cases contamination can produce irreversible effects. From a biological stand, water quality is conditioned by the absence or presence of living organisms in the aquatic ecosystem, as well as by the greater or lesser presence of pathogenic agents.

Water quality in the Copaquilla trout farm (**Table 1**) is within normal ranges. Exceptional were some parameters in specific conditions, outside the optimal range, with no result in harming the crop itself.

In general, water used for trout farming complies within Chilean standard NCh1333.Of78 Mod. 1987, regarding the maximum limits allowed for water used in aquatic life farming.

7. Acquisition and transport of rainbow trout specimens

Trout were purchased at Rio Blanco fish farm, a department of Universidad Católica de Valparaíso, located in the city of Los Andes, Valparaíso Region.

The truck used to transport the trout was equipped with a thermos that facilitated the regulation of the internal environmental temperature during transport. A support vehicle was guarding it in case of emergency. These vehicles were disinfected (**Figure 4**) before entering Rio Blanco fish farm, in accordance with the fish transport procedures established by Servicio Nacional de Pesca y Acuicultura (SERNAPESCA) in Chile (or National Fisheries and Aquaculture Service).

Once the truck arrived at Rio Blanco fish farm, 8 ponds of 1 m3 each inside of it were filled with water (**Figure 5**). The ponds housed the fish during the transfer, being loaded at a rate of 625 fish per tank. Once the loading was completed, the outlet temperature and oxygen were recorded, and the ponds were carefully closed.

Parameters mentioned previously were taken (**Figure 6**) during the first 6 hours of the trip and visual evaluation was carried out by a technical team in charge of the transport. Behavior of the animals was evaluated, mainly swimming and opercular movement. During the following 12 hours, parameters were measured, and visual evaluation was carried out every two hours. Twelve more hours later, parameters were taken every three hours. For the last 15 hours they were measured every 5 hours. Thus, completing a total of 45 hours of travel.

Once the fish arrived at the top of Copaquilla, they were transferred to 800-liter water ponds on a pickup truck that carry them to the farm. Eight trips were made to empty the eight ponds. Once all the fish were in the farming center, they were distributed in three 40 m^3 ponds in equivalent number of fish.

Figure 4.
Disinfection of truck and support vehicle, for rainbow trout transport.

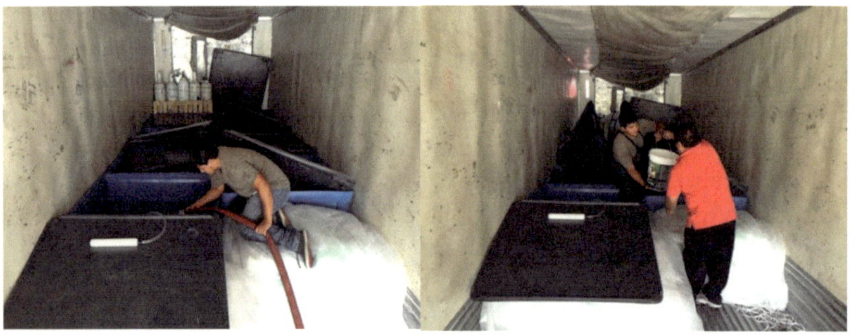

Figure 5.
Filling of water and fish to be transported into the ponds.

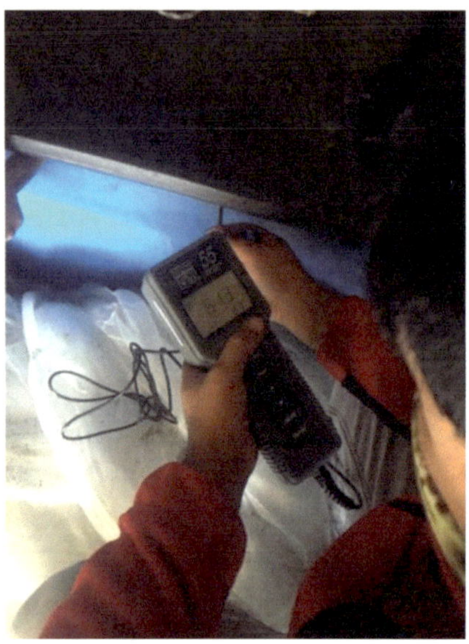

Figure 6.
Measurement of parameters during the transport of rainbow trout.

8. Development of the trout farming system

8.1 Trout fattening

A large part of the productive efficiency of a commercial fish farm is determined by feed management. For it to be successful, it is essential that farms develop a database where the indispensable parameters for a correct management

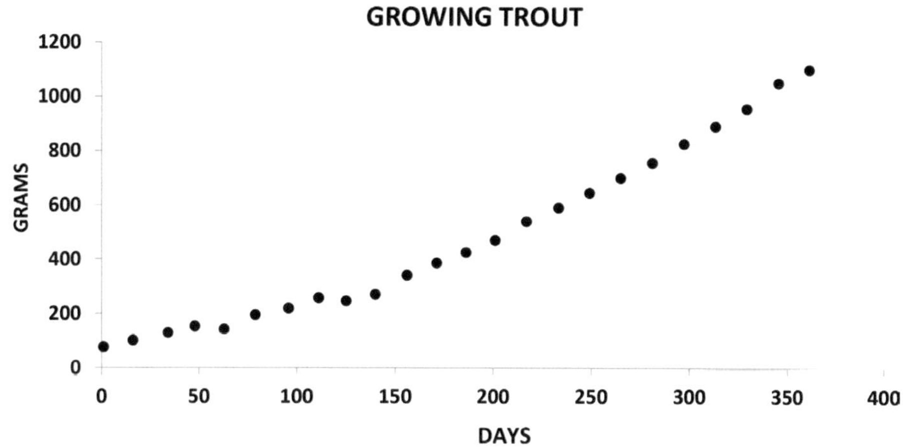

Figure 7.
Rainbow trout growth in the CCPC.

of feeding are recorded. Among these it is worth mentioning; number of individuals corresponding to each pond, mortality, average body weight, growth evolution, feed supply and water temperature. Once the numbers are in, it is possible to calculate the ration to be supplied to each pond as well as the feeding efficiency of each one.

The fish were sampled every 15 days. Between 5 and 7% of the total fish per pond were sampled, in this way, it is possible to correct the variables used to calculate the feed ration to be supplied month by month. The number of fish in the ponds was calculated as the difference between the initial number of the crop minus the number of dead individuals. Mortality was recorded daily to keep numerical control of the fish. With the updated number of individuals per pond and the average body weight records, the total biomass of the crop could be determined (**Figure 7**). After several samplings, the growth rate of the fish could be calculated by the difference in average weights, allowing for definition of trout harvest.

For an optimized determination of the feed ration supplied to the trout, the biomass, water temperature and average individual body weight in the pond must be known. Initially, the commonly disseminated feeding tables for the species can be used as a guide. However, we elaborate our own feeding table under specific conditions of the crop. Under this concept the fish were fed four times a day at a rate of 3 to 4% of the fish biomass in the pond. The conversion factor fluctuated between 1.2 and 1.35.

According to general results obtained in the present study, it is not recommended to apply any type of food restriction when the trout are growing within the optimum temperature range for the species, i.e., they were fed *ad-libitum*.

8.2 Conditioning of breeding specimens at CCPC

After one year of farming, 30 specimens were selected out of the 6,000 15-gram fish that initially arrived. These were separated in a 40 m^3 pond as potential broodstock. A density of 1 kilo of fish per m^3 was considered, feeding was supplied three times per week, twice with trout broodstock pellet and once with normal fattening pellet. Water flow was 120 liters per minute.

Fish were selected mainly for their phenotypic appearance, rapid growth rate and presenting no deformities. After one year they had an average weight ranging from 1,300 to 1,500 grams.

8.3 First trout spawning and hatching at the CCPC

In August 2017, the trout had their first spawning. This process was carried out through an abdominal massage on the female, which expelled her oocytes over a dry stainless-steel container. Already having the females spawn, the males were similarly massaged (two to three males per female). Once the gametes were in the container, they were activated by introducing clean water, which was also used as a mean to wash the eggs. This process was carried out several times until the water turned completely transparent. The eggs were then placed in trays adapted for incubation (**Figure 8A**) in raceway-type ponds of approximately 550 liters. After the first 48 hours, the eggs in the trays were cleaned, removing dead eggs in the process (**Figure 8B**) and finally leaving them unhandled until the eyed stage [10].

About 15,000 eggs were fertilized and hatched, reaching approximately 10,500 eggs to larvae (**Figure 9**), 6,500 to juveniles and 2,400 to 160-gram adults as of June 2018. It should be noted that a large percentage of animals were removed at each stage due to the limited space available at the CCPC.

Water temperature in the hatchery trays varied between 7 and 10 degrees at night and between 12 and 15 degrees during the day. The water flow in the raceway-type ponds was 5 liters per minute. For each pond there were four trays that held the eggs. Each tray was provided with a wire mesh in front of the water flow and at the bottom of the tray, which allowed water circulation and oxygenation.

Once the eggs hatched in the trays and the larvae absorbed the yolk sac, they were transferred directly to the raceway ponds, where they were kept for 20 to 25 days (**Figure 10**).

A B

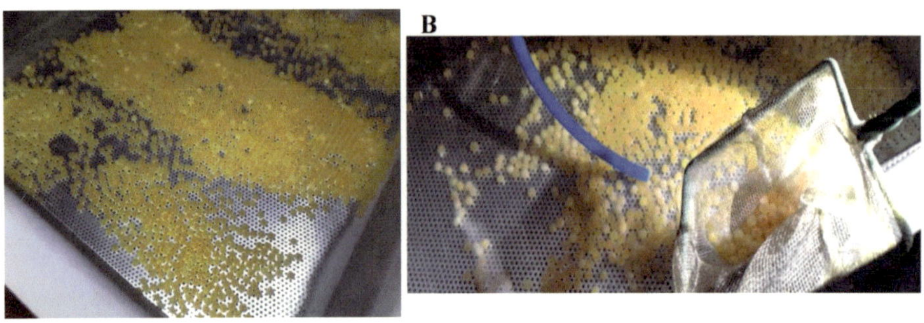

Figure 8.
Trout eggs (A), extraction of dead eggs (B).

A B

Figure 9.
Eggs in hatchery trays (A), rainbow trout larvae in the CCPC (B).

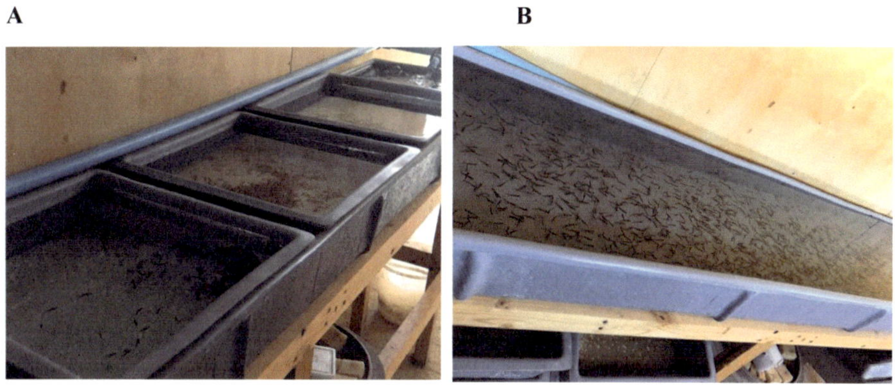

Figure 10.
Raceway pond, hatchery trays A, juveniles B.

Figure 11.
Trout fry ponds.

Figure 12.
Trout fattening ponds of 40 mt^3.

Temperature and water parameters were maintained during the time juveniles were in these ponds. Once the stage in the raceway ponds was completed, juveniles were transferred to the 450-liter fry tanks (**Figure 11**). They remained there for a month, approximately.

This system had a completely independent SAR, with a 0.5 hp. pump, 0.2 hp. blower and 80 w UV lamp. Temperature parameters fluctuated between 9 and 12°C at night and 12 and 16°C during the day. Water flow was 8 to 10 lt m^{-1}. After thirty days, the fish were transferred to a 40 m^3 tank (**Figure 12**). They were to remain there until they reached 160 grams, at which point the study was completed and splitting and grading were performed.

9. Problems addressed in trout farming

Trout farming in the Copaquilla sector was not exempt from problems inherent to fish farming and the climatic conditions of a sector located at 3,000 mamls.

Immature biofilter, massive proliferation of microalgae and accumulation of organic matter at the bottom of the water storage pond, were recurrent problems at the beginning of the culture.

They led to rapid responses in order to ensure the survival of the fish. One of the most worrying natural factors were summer rains, known as "the altiplanic winter". The natural phenomenon causes rains and cloudy days that affect the efficient operation of the solar panels, a system on which the energetic exercise of the equipment (pumps, blower, oxygen generator, etc.) depends.

10. Prospects for viability and sustainability of the Copaquilla Pukara farming center

The implementation of an aquaculture system in the pre-Cordilleran region made it possible to disseminate knowledge about it and to predict its viability under the surrounding conditions (quality of the water). In order to implement this type of technology on a large scale in the future, both for trout farming systems and for other species, the first step had to be made. Therefore, it is a means by which companies and entrepreneurs can obtain investment resources that allow for the beginning of new companies. This is based off the need to diversify aquaculture and the region's need to promote it; a strategic axis for the region of Arica and Parinacota. In addition, this aquaculture technology is highly suitable for coupling it with solar energy, as it has been demonstrated at the CCPC.

This initiative also poses a potential for entrepreneurship and innovation through the implementation of aquaculture and training of personnel in the region. Newly trained personnel would be capable of operating the crops, analyzing water quality, applying growth rates, among other activities. This, in turn, will allow new studies to be carried out, complementing the ones related to economic activities in which aquaculture plays an important role. The implemented equipment will allow the follow through continuum of the procedures, generating positive externalities such as internships for high school and university students.

Finally, the CCPC is an important business opportunity for small companies in the region that consider aquaculture in surrounding communities and for the sustainability and permanent operation of aquaculture production. In general, the CCPC will generate a preponderant added value that will help to promote and strengthen small businesses. Regarding environmental mitigation measures and risk prevention, the project is entirely sustainable from an environmental stand, as it allows water to be continually reused. Furthermore, this productive activity is intended to prevent the migration of young people from their native villages, providing them with a work alternative and specialization in aquaculture. Lastly, the implementation of the farm and other equipment does not create a significant visual impact as it is installed in a small area similar to local structures.

Author details

Renzo Pepe-Victoriano[1*], Héctor Aravena-Ambrosetti[1] and Piera Pepe-Vargas[2]

1 Área de Biología Marina y Acuicultura, Facultad de Recursos Naturales Renovables, Universidad Arturo Prat, Arica, Chile

2 Facultad de Ciencias Naturales y Oceanográficas, Universidad de Concepción, Concepción, Chile

*Address all correspondence to: rpepev@unap.cl

References

[1] Pepe-Victoriano, R. M. Araya, C. Wurmann, J. Mery, A. Vélez, P. Oxa, C. León, S. Fuentealba & R. Canales. 2015. Estrategia para el desarrollo de la Acuicultura en la región de Arica y Parinacota, 2015-2024. Informe Final CORFO. 453 pp.

[2] Pepe-Victoriano, R & H. Aravena-Ambrosetti. 2018. Primer Centro de Cultivos de Truchas arcoíris bajo un sistema de Recirculación en el sector de Copaquilla, Región de Arica y Parinacota. Rev. Versión Diferente 15: 58-62.

[3] Timmons, M. J. Ebeling, R. Piedrahita. 2009. Acuicultura en Sistemas de Recirculación. NRACE Publication Nº 101-209 Spanish. 959 pp

[4] Food and Agriculture Organization (FAO), 2016. El estado mundial de la pesca y la acuicultura. Organización de las Naciones unidas para la Agricultura y la Alimentación, Departamento de Pesca y Acuicultura, Rome.

[5] Cornejo-Ponce, L., Vilca-Salinas, P., Lienqueo-Aburto H., Arenas, M., Pepe-Victoriano, R., Carpio, E., and Rodriguez J. 2020. Integrated Aquaculture Recirculation System (IARS) Supported by Solar Energy as a Circular Economy Alternative for Resilient Communities in Arid/Semi-Arid Zones in Southern South America: A Case Study in the Camarones Town. Water. MDPI. 12, 3469; doi:10.3390/w12123469

[6] Gall G & Crandell P (1992) The rainbow trout. Aquaculture 100, 1-10

[7] Groot C (1996) Chapter 3 Salmonid life histories. Developments in Aquaculture and Fisheries Science 29, 97-230.

[8] Laird L & Needham T (1988) The farmed salmonids. In: L.M. Laird and T. Needham (Editors), Salmon and Trout Farming. Ellis Horwood, Chichester 15-31.

[9] Espinós A (1995) El alimento - presa que encontraron las truchas en la Argentina Irene Wais. Boletín mosquero AAPM invierno de 1995. Los salmónidos (enfoque científico) APMN Cornell University - Department of Natural Resources Novascotia - Canada - Agriculture and Fisheries

[10] Blanco, M. 1995. La Trucha, Cría Industrial. Ediciones Mundi-Prensa, Madrid-España. 503. pp.

Chapter 3

Developments in Probiotic Use in the Aquaculture of *Salmo* Spp.

Alexander Dindial

Abstract

While interest in probiotic use in aquaculture is not a new phenomenon, the past few years have seen great developments in probiotic research in *Salmo* spp.. This review examines the corpus of literature surrounding the use of probiotics in some of the species of *Salmo* most important to modern aquaculture, including *Salmo salar*, *S. coruhensis*, *S. trutta,* and *S. trutta caspius,* with a particular emphasis on the most recent research. The use of many of these probiotics is associated with such host benefits as enhanced growth, nutrition, and immunity. These benefits and the potential applicability of these probiotics to the modern aquaculture of *Salmo* are reviewed herein.

Keywords: Salmo, Atlantic salmon, brown trout, Coruh River trout, Caspian brown trout, *S. salar*, *S. trutta caspius*, *S. coruhensis*, *S. trutta*, fish probiotics, aquaculture probiotics

1. Introduction

As in other animals, the gut microbiome of fish is a dynamic, complex, organ-like system that may contain trillions of microorganisms. It is implicated in a number of functions critical to the animal's health, including digestion, immunity, and nutrient absorption, which may in turn contribute to such things as development, growth and metabolism [1].

There are a few ways by which these salubrious phenomena are known to occur. In terms of immunity, some beneficial bacteria may produce such things as bacteriocins, peroxides, and acids in order to inhibit the growth of pathogenic microorganisms, and have further been observed to mitigate damage caused to the intestines by pathogens [2, 3]. Furthermore, some bacteria have been known to promote the host fish's immunity through such means as the enhanced infiltration of epithelial leukocytes, the modulation of cytokine and chemokine expression, enhanced leukocyte activity, and enhanced lysozyme activity in the cells of the mucosa [4–8]. On the other hand, some bacteria have been known to enhance fish digestion and nutrient absorption through such things as increasing the length of microvilli, enhancing the fold length of the mucosa, secreting various digestive enzymes, and by fermenting non-digestible compounds (e.g., complex carbohydrates) to make them usable [4, 9–11].

In order to take advantage of some of these beneficial health effects, it is critically important that fish reared in aquaculture maintain healthy gut microbiota. This may help ensure the success of aquaculture operations by preventing the spread of disease and promoting the healthy growth of fish to market size. To this end, there has been a considerable amount of research on the fish microbiome and the use of probiotics in fish significant to the field of aquaculture. In general, the term "probiotic" typically refers to a sample of live microorganisms that is used in order to confer some kind of health benefit [12]. In many cases, probiotics are consumed in order to allow the microorganisms to enter the organism's gut, from which point they can exert their salubrious effects. In some cases, probiotics are combined with prebiotics, which are compounds designed to bolster the growth of beneficial bacteria. In many cases, these oligosaccharides can be digested by the bacteria, but not the host. The combination of probiotics and prebiotics is sometimes referred to as synbiotics [12].

Of the aquaculturally significant fish in which probiotics have been studied, genus *Salmo* is perhaps among the most important. This genus contains a number of species relevant in modern aquaculture, including *Salmo salar* (the Atlantic salmon), *Salmo trutta* (the brown trout), *S. trutta caspius* (the Caspian trout/salmon) and *S. coruhensis* (the Coruh River trout), *inter alias*. Of these species, *S. salar* is perhaps the most important to global aquaculture. According to the Food and Agriculture Organization of the United Nations, farmed *S. salar* comprises greater than 90% of all farmed salmon and greater than half of all salmon production worldwide [13].

Underscoring the importance of *Salmo* in modern aquaculture, there is a rapidly growing corpus of literature surrounding the use of probiotics in this genus. The benefits to fish immunity and growth that stem from this research may have the potential to greatly benefit the salmon aquaculture industry, thus necessitating further review of the recent pertinent literature.

1.1 Overview of the gut microbiota of *Salmo spp.*

To understand the use of probiotics in *Salmo*, it is important to first critically examine this genus' gut microbiota. The intestinal mucosa of *S. salar* can harbor trillions of microorganisms, the majority of which are bacteria. Typically, the identity of these microorganisms may vary considerably depending on several variables, including diet, the bacteria present, biogeography and environmental factors, stress, captivity, disease, the host's species, and life cycle stage [14–19]. To further complicate matters, not all microorganisms in the salmon gut are permanent residents. While many bacteria are indeed capable of colonizing the intestinal mucosa in the long-term, the presence of others may only be transient [20]. Such transience may be due to a number of factors, including the inability to compete with bacteria that have already colonized the gut for such things as nutrients or space, direct inhibition by pre-existing bacteria (e.g., via the secretion of antimicrobial peptides), or even just the general inability to colonize the intestinal tract [2, 21].

Regardless of this dynamism and complexity, researchers have identified a diverse array of bacteria that may inhabit the digestive system of *Salmo*. In one study, the mid and distal intestinal mucosa of *S. salar* kept in seawater were found to contain bacteria from a number of phyla, including Proteobacteria (which comprised ~90% of all mucosal bacteria), Actinobacteria, Armatimonadetes, Spirochaetes, Bacteroidetes, and Firmicutes [20]. Furthermore, in the digesta of *S. salar*, Proteobacteria, Firmicutes, Fusobacteria, and Bacteroidetes (among other less abundant bacteria) were documented [20]. In another study, the digesta of *S. salar* parr reared in a freshwater loch or a recirculating aquaculture system (RAS) were also found to contain representatives of other additional phyla, including Tenericutes and Acidobacteria, among others [22]. In a different study featuring wild *S. salar* at different life cycle stages, Nitrospirae were also found to be abundant in the digestive tracts of parrs, smolts, and adults. This study also found that

Firmicutes, Actinobacteria, and Bacteroidetes occur at far lower levels in marine adults than in freshwater life stages [18]. Overall, these phyla are represented by dozens of genera in the *S. salar* microbiome, notably including *Carnobacterium*, *Lactobacillus*, *Pediococcus*, *Lactococcus*, *Vibrio*, *Pseudomonas*, *Aeromonas*, *Yersinia*, and *Mycoplasma*, many of which will be discussed further in this review [18, 20].

2. Probiotic use in *Salmo*

In general, colonization is critically important to the establishment of probiotic bacteria within a fish's digestive tract. There are a few main considerations behind this probiotic colonization. Firstly, the bacteria must not pose any danger to the host fish or be otherwise pathogenic. Instead, they should exhibit salubrious effects as described previously. Secondly, it is critical that the bacteria are able to reproduce successfully within the fish's gut, such that their rate of multiplication exceeds the rate at which they are expelled [2]. Thirdly, the bacteria must be able to adhere to the intestinal mucosa. There are a few additional considerations to this third point. For example, the bacteria must be able to successfully compete with other bacteria for adhesion space. Some bacteria (including some lactic acid bacteria) are known to have specific adhesion sites on the intestinal epithelium, while others are known to adhere non-specifically [2, 23, 24]. These considerations are critically important to all potential fish probiotics.

Table 1 provides an overview of the bacterial genera that have been investigated as probiotics in genus *Salmo*:

***Salmo* species**	**Bacterial genus**	**Bacterial species**	**Citation**
Salmo trutta caspius	*Bacillus*	*B. subtilis & B. licheniformis**	[25–27]
	Pediococcus	*P. acidilactici***	[28–30]
	Lactobacillus	*L. plantarum*	[31]
Salmo salar	*Carnobacterium*	*Carnobacterium* spp.	[9]
		C. divergens	[3, 32]
	Pediococcus	*P. acidilactici***	[11, 33]
	Lactobacillus	*L. delbrueckii*	[34]

Asterisks denote the use of commercial probiotic formulations: one asterisk denotes BetaPlus ® and two asterisks denote Bactocell ®. These products are discussed in greater detail below. See ***Table 2*** *for an overview of other probiotic formulations.*

Table 1.
An overview of different bacterial species investigated as probiotics in genus Salmo.

2.1 *Carnobacterium*

Carnobacterium is a genus of Gram-positive lactic acid bacteria (LAB) within the phylum Firmicutes. They are ubiquitous in nature and can survive low temperatures and anaerobic conditions with elevated concentrations of carbon dioxide [35]. In terms of fish, some *Carnobacteria* (such as some strains of *Carnobacterium maltaromaticum*) have been known to be pathogenic in certain salmonids (e.g., *Oncorhynchus mykiss*), while others have been found to exhibit beneficial effects within the gut microbiome [9, 32, 35, 36]. Perhaps one of the most important benefits of probiotic *Carnobacteria* is their ability to inhibit the growth of pathogenic bacteria within the *Salmo* gut. This ability is most likely due to *Carnobacteria's* ability to produce antimicrobial bacteriocins, which may serve to both inhibit pathogenic bacteria and help the *Carnobacteria* to survive within the competitive environment of the gut microbiome [37]. However, the ability of probiotic *Carnobacterium* to inhibit different pathogenic species is known to vary by strain [32, 37]. As a type of LAB, *Carnobacteria* are also capable of lactic acid production, which may also have an inhibitory effect on pathogens. Further recent research has also suggested that the presence of *Carnobacterium* in the pyloric caeca of *S. salar* is associated with enhanced flesh color. While it is hypothesized that this may be related to the production of carotenoids by the *Carnobacteria* as well as their pro-immune effects, the true reason for this phenomenon remains unclear [38].

In an early study conducted in 2000, it was revealed that *Carnobacterium* strains isolated from the intestine of *S. salar* could inhibit a number of pathogenic bacteria, including *Aeromonas hydrophila, Aeromonas salmonicida, Vibrio ordalli, Vibrio anguillarum, Streptococcus milleri, Photobacterium damselae piscicida,* and *Flavobacterium psychrophilium in vitro* [9]. In this study, it was also found that the administration of these *Carnobacteria* for at least fourteen days was able to promote the survival of *O. mykiss* in the context of infection by *A. salmonicida, Yersinia ruckerii,* and *V. ordalli.* This study also found that it took 28 days of probiotic administration to achieve the maximum intestinal *Carnobacterium* levels, and that the cessation of probiotic administration in fry and fingerlings resulted in the bacteria becoming undetectable in the gut in ten days or less [9]. It was further confirmed in another study the coincubation of *C. divergens* strain 6251 with the pathogenic *A. salmonicida* and *V. anguillarum* was able to prevent (but not alleviate) damage to *S. salar* microvilli [3]. A later study demonstrated that the administration of a commercial prebiotic (namely EWOS prebiosal) to help promote probiotic bacterial growth vastly enhanced the ability of *C. divergens* to adhere to the epithelia and mucosa of the proximal (but not distal) intestine of *S. salar* [32].

Overall, certain strains of genus *Carnobacterium* (such as the aforementioned *C. divergens* strain 6251) may have great potential as a probiotic for *S. salar*, with the ability to bolster the fish's immunity and perhaps even flesh color. While it is possible that the probiotic may need to be frequently readministered, the use of a commercially available prebiotic may be beneficial in enhancing bacterial adhesion to the salmon gut.

2.2 *Pediococcus*

Pediococcus is a genus of Gram-positive LAB within the phylum Firmicutes. It is an acidophilic, facultative anaerobe with well-established probiotic properties, even in humans. Like *Carnobacterium*, *Pediococcus acidilactici* is known to produce bacteriocins and lactic acid, which are potentially useful in inhibiting the propagation of pathogens within the digestive tract [39].

P. acidilactici has been investigated as potential probiotic in both *S. salar* and *S. trutta caspius.* In both fish, much of this research has been conducted using Bactocell ®, a commercially available strain of *P. acidilactici* (strain MA 18/5 M) that is also used in other animals of agricultural significance [28–30]. It is one of the only aquacultural probiotics approved in the European Union [33]. Overall, Bactocell ® seems to exhibit some promise in promoting the immunity and growth of *S. trutta caspius*. In one study, Bactocell ® was found to significantly decrease the feed conversion ratio in *S. trutta caspius* following five treatments with the probiotic. Furthermore, white blood cell concentrations were noted to have increased, suggesting pro-immune effects related to the presence of the probiotic. Curiously, red blood cell counts were also found to be lowered relative to the control group that did not receive the Bactocell ® treatment [30]. In another study in which Bactocell

® was co-administered with iron, parameters like body weight gain and specific growth rate were also found to be enhanced relative to the control group [28].

This same strain of *P. acidilactici* (MA 18/5 M) has also been investigated as a probiotic in *S. salar.* In one such study, saltwater Atlantic salmon were administered *P. acidilactici* MA 18/5 M and short chain fructooligosaccharides (as a prebiotic) twice a day for 63 days. Among other things, the researchers noted that this synbiotic (a combination of a probiotic and prebiotic) modulated both local and systemic immunity. For example, it was found that synbiotic administration was associated with the increased expression of the pro-inflammatory cytokines interleukin-1β, tumor necrosis factor α, and interleukin in the intestinal tissue, in addition to increased expression of the antiviral molecules toll-like receptor 3 and myxovirus-resistant protein 1. Furthermore, the researchers observed increased epithelial leukocyte infiltration in the intestines and elevated serum lysozyme levels. Notably, the fish that received the synbiotic also exhibited greater villus length relative to the control and decreased but recoverable intestinal bacterial loads (without adverse health consequences) [11]. In another study, it was found that Bactocell ® was capable of modulating several parameters related to distal intestine inflammation, which overall helped to counteract the inflammation [40]. A later study further studied the administration of *P. acidilacti* MA 18/5 M to *S. salar*, in addition to also considering the effects on the gut microbiome that occur with the transition from the freshwater and saltwater stages. Overall, *Pediococcus* was found to significantly impact the Atlantic salmon microbiota, as in the previous study. Notably, among the fish that received the probiotic treatment, *Pediococcus* was present in greater abundance in both the digesta and mucosa in the freshwater salmon than in the saltwater salmon. In spite of this, the effect of the probiotic treatment on the composition of the gut microbiota was found to be the greatest among the saltwater fish [33].

In summary, there is some evidence to suggest that *P. acidilactici* MA 18/5 M may exhibit positive effects on the growth of *Salmo* spp., as well as significant immunomodulatory effects. Furthermore, its availability on the market may make this strain an enticing choice for those interested in using effective probiotics in their aquaculture systems.

2.3 *Bacillus*

Bacillus is a diverse genus of Gram-positive bacteria within the phylum Firmicutes. There are three main species of *Bacillus* that have been investigated as probiotics for *Salmo*, the two most notable of which are *B. subtilis* and *B. licheniformis*. While *Bacillus* species are not LAB, they have nevertheless been successfully implemented as probiotic agents in aquaculture, exhibiting the ability to enhance growth and immunity in a few species [41, 42].

In terms of *Salmo*, *B. subtilis* DSM 5749 and *B. licheniformis* DSM 5750 have been investigated as probiotics in *S. trutta caspius* in the form of the commercially available product known as BetaPlus ®. In one study, Caspian salmon fingerlings were administered a synbiotic composed of BetaPlus ® and galacto-oligosaccharides. In comparison to the control (which did not receive the synbiotic), the group that received the synbiotic exhibited superior performance in parameters relevant to growth and immunity (among other things). In terms of growth, this includes a lower feed conversion ratio and higher weight gain and protein efficiency ratios. As for immunity, the fish that received the probiotic exhibited increased serum levels of lysozyme, immunoglobulins, bactericidal peptides, agglutinins, lectins, and albumin [42]. In another, similar study with the same species, BetaPlus ® was used in conjunction with isomaltooligosaccharides as the prebiotic. In addition to some similar findings to the previous study, the synbiotic group was found to exhibit greater levels of monocytes, leukocytes, and neutrophils compared to the control, which is further suggestive of enhanced immunity in the context of synbiotic usage [25].

Overall, these findings suggest that the administration of BetaPlus® in conjunction with a prebiotic may be able to significantly enhance both growth and systemic immunity in *S. trutta caspius*. For those involved in the commercial aquaculture of *Salmo* spp., it may be beneficial to consider the use of BetaPlus® or similar probiotic formulations.

2.4 *Lactobacillus*

Lactobacillus is a genus of Gram-positive LAB within the phylum Firmicutes. Two species in this genus have been investigated as probiotics for species within *Salmo*, namely *Lactobacillus delbrueckii* in *S. salar* and *L. plantarum* in *S. trutta caspius.*

The use of *Lactobacillus delbrueckii lactis* as an aquacultural probiotic has been known to enhance the innate immune response in other fish, such as *Sparus aurata* [43]. In terms of the Atlantic salmon, one study examined this bacterium's ability to remain on the surface of the intestines in an *in vitro* model containing intestinal tissue from an Atlantic salmon, as well as its ability to prevent damage in the context of an *Aeromonas salmoncida salmoncida* infection. Overall, it was found that the *Lactobacillus* was able to persist on the surface of the intestine, and further caused no damage to the tissue (in contrast to the *Aeromonas*). When the *in vitro* intestine model was co-incubated with both the *Lactobacillus* and *Aeromonas*, the former prevented damage to the tissue caused by the latter [34]. Overall, this is likely suggestive of the ability of *L. delbruecki lactis* to contribute to host innate immunity in *S. salar.*

Further work in *S. trutta caspius* provides some evidence for the ability of *L. plantarum*-based synbiotics to promote both host growth and immunity. In one study, Caspian salmon were assigned into eight groups, which featured combinations including the fishes' basal diet, *L. plantarum*, and the prebiotics beta-glucan and mannan oligosaccharide. All groups featuring probiotics and/or prebiotics exhibited decreased feed conversion ratios and feed intake, as well as increased weight gain and protein efficiency ratios. These same groups also exhibited enhanced parameters relevant to immunity, including elevated levels of immunoglobulin M, and lysozyme (*inter alia*). It is also important to note that with the exception of the beta-glucan group, the groups featuring *L. plantarum* exhibited lower cortisol and glucose levels than the other experimental group, suggesting yet another physiological benefit of the use of *L. plantarum* as a probiotic [31].

In toto, while there are not many available studies focusing on the use of *Lactobacillus* spp. as a probiotic in *Salmo* on their own, the promising findings related to growth, immunity, and decreased feed intake associated with this bacterial genus may warrant further investigation as a probiotic in *Salmo.*

2.5 Other probiotics

In addition to the probiotics discussed above, there are some other probiotics that have been investigated in *Salmo*, including those with multi-genus or variable composition:

2.5.1 Kefir

There has been some research on the use of kefir as a probiotic for *S. coruhensis*. Kefir is a fermented, dairy-based beverage with origins in the North Caucasus and has been used as a probiotic in humans. Kefir is known to contain a variety of different microorganisms, including representatives of *Lactobacillus, Lactococcus,* and *Leuconostoc*, as well as other LAB, acetic acid bacteria, and even yeasts [44–46]. Overall, the use of kefir as a probiotic in this species seems to offer some promise. In one study, it was demonstrated that groups of Coruh trout administered kefir exhibited decreased activity of catalase, an enzyme with antioxidant activity, as well as decreased levels of the highly reactive compound malondialdehyde in hepatic tissue. This may suggest that the probiotics associated with kefir may have antioxidant activity [44]. In another study with the same species, kefir administration at a dose of ten or twenty grams per kilogram was associated with elevated immunoglobulin, which may suggest that kefir's probiotics may also exhibit immunomodulatory effects in the Coruh trout. However, this same study did not find any difference in growth or survival rates between the control and experimental groups [45]. Finally, in another study, it was found that Coruh trout in the groups administered kefir also exhibited changes in digestive and hepatic enzyme expression and glucose levels relative to the control group. For example, it was found that serum amylase and lipase levels, as well as serum glucose decreased, suggesting that kefir administration may modulate digestion in *S. coruhensis*. Notably, glucose levels were also found to be lower in the kefir groups, which may be suggestive of decreased stress or decreased carbohydrate absorption in the intestines [46].

Overall, kefir may show some promise as a probiotic for *S. coruhensis*, especially considering the relative ease by which it may be acquired and its antioxidant and immunomodulatory properties. However, its inability to affect host growth or survival rates (unlike other previously discussed probiotics) may warrant careful consideration when choosing it as a probiotic for *Salmo* spp..

2.5.2 Other multi-species probiotic formulations

Other probiotic formulations featuring multiple species of microorganisms have also been investigated. One notable example of such a formulation is Bio-aqua ®, an aquacultural probiotic that is commercially available in Iran. In addition to the species enumerated in **Table 2**, Bio-aqua® further contains yeast extract and fructooligosaccharides as prebiotics. In the sole English language publication investigating this formulation, it was found that Bio-aqua did not produce any significant effects on the growth performance, activity of digestive enzymes, or intestinal histomorphology in juvenile *S. trutta caspius* when administered at a dose of 0.2 grams per kilogram at feeding times [50].

In another study, four experimental groups of juvenile *S. salar* reared in a RAS were administered combinations of *Rhodotorula mucilaginosa* CGMCC 1013 and *Bacillus velezensis* V4 CGMCC 10149 over a 62-day period. Relative to the control group, the experimental group fish exhibited significantly decreased feed conversion ratios and mortality, as well as increased weight gain ratios and specific growth rates. Furthermore, immunological and antioxidant parameters were suggestive of an enhanced immune response and antioxidant activity in the experimental groups in comparison with the control group. Curiously, however, cortisol levels were found to be elevated in the experimental groups relative to the control. Finally, in a challenge trial conducted with *Aeromonas salmonicida*, the experimental group fish exhibited far lower mortality rates than the control group, perhaps due to a combination of the enhanced immune response and the direct inhibition of the pathogen by the probiotic bacteria in the gut [51].

Overall, in spite of the lack of observed benefits in the study featuring Bio-aqua ®, the promising findings of the second study featuring *R. mucilaginosa* and

Salmo species	Probiotic formulation	Probiotic microbes	Citation
S. coruhensis	Kefir	*Lactobacillus*, *Lactococcus*, *Leuconostoc*, other lactic or acetic acid bacteria, and yeasts**	[44–46]
S. salar, *S. trutta*, & *S. trutta caspius*	Fermented soymeal***	Variable, but may include lactic acid bacteria, acetic acid bacteria, and/or yeasts**	[21, 47–49]
S. trutta caspius	Bio-aqua ® *	*Pediococcus acidilactici*, *Enterococcus faecium*, *Bacillus subtilis*, *Lactobacillus acidophilus*, *Lactobacillus plantarum*, *Lactobacillus casei*, *Lactobacillus rhamnosus*, *Bifidobacterium bifidum*, *Saccharomyces cerevisiae***	[50]
S. salar	N/A	*Bacillus velezensis*, *Rhodotorula mucilaginosa*	[51]

One asterisk signifies a commercial probiotic formulation, whereas two asterisks signify a non-bacterial microorganism. Three asterisks denote a hypothetical probiotic source.

Table 2.
An overview of probiotic formulations featuring multiple bacterial species that have been investigated for use in Salmo.

B. velezensis suggest that the utilization of multiple species of bacteria in probiotics has the potential to be highly productive.

2.5.3 Fermented soybean meal

There has also been considerable research on the use of fermented plant meals (such as soybean meal) in the aquaculture of *Salmo,* including *S. salar, S. trutta,* and *S. trutta caspius* [47–49]. However, it is important to note that this usage of plant meal fermentation is often implemented to enhance the bioavailability of nutrients and energy in plant products that are not natural components of the salmonid diet [52]. Regardless, it is possible that some fermented plant meal products may exhibit probiotic or prebiotic effects in *Salmo.* In one study, the provision of fermented soybean meal to *S. salar* was found to be associated with an increase in the abundance of intestinal LAB (including *Lactobacillus, Pediococcus,* and *Lactococcus*) when compared with a group that was fed a fishmeal-based diet and another that received non-fermented soybean meal [21]. While it is unclear from the results of the study whether this increase in LAB abundance was due to prebiotic or probiotic effects, it is well-established that certain fermented foods (such as miso, which is also soybean-based) are associated with probiotic LAB populations [53, 54]. Therefore, it is plausible that at least some fermented soybean feeds for *Salmo* may exhibit probiotic effects. However, further research may be needed in order to investigate this hypothesis.

3. Conclusions

In general, the use of probiotics in the aquaculture of species within the genus *Salmo* has the potential to be highly productive, with the ability to promote fish immunity, nutrition, and growth as well as to decrease the likelihood of mortality. These beneficial effects are more important now than ever, with the increasing popularity of intensive aquaculture systems like RAS. In such crowded conditions, fish are constantly exposed to pathogens, underscoring the need for enhanced immunity. Furthermore, the enhanced growth and decreased mortality from certain diseases associated with the use of some probiotics may help the operators of salmon aquaculture systems to increase their profits and minimize the unnecessary loss of fish. The current market availability of such tested probiotic formulations as BetaPlus ® and Bactocell ® may also be instrumental in helping salmon aquaculturalists to achieve these outcomes.

In spite of the benefits that can be understood from the current corpus of literature pertaining to the use of probiotics in *Salmo*, many questions remain unanswered. Some of these questions are quite simple, and could consider such things as which probiotic species and strains (or perhaps combinations thereof) are the best at achieving particular outcomes in particular species of *Salmo* (e.g., which species and strains result in optimal growth rates in *S. salar*). The answers to such questions would likely necessitate the conduction of relatively large studies featuring multiple experimental groups that each receive different probiotics. Other questions concern the biochemical activity of probiotic bacteria within the host, including such things as how or why certain probiotics influence the host's immune system (e.g., how does BetaPlus® cause elevated white blood cell levels in *S. salar* or, why does the administration of certain probiotics result in elevated immunoglobulin levels). It further remains unclear as to why some probiotic species and strains differ from others in certain effects on the host (e.g., why were host cortisol levels decreased relative to the control when *Lactobacillus plantarum* was used in *S. trutta caspius,* but elevated when *Bacillus velezensis* and *Rhodotorula mucilaginosa* were used in *S. salar*). The elucidation of answers to these (and related) questions may help us better understand the relationship between the salmon gut microbiota and fish health, as well as potentially inform future efforts to optimize the salmon gut microbiome with biotechnology.

Overall, while much remains to be explored in the use of probiotics in *Salmo*, recent findings have strongly indicated that they are associated with remarkable potential benefits that should warrant any *Salmo* aquaculturalist to consider their use.

Author details

Alexander Dindial
Zanvyl Krieger School of Arts and Sciences, Johns Hopkins University, Baltimore, MD, USA

*Address all correspondence to: adindia1@jh.edu

References

[1] Talwar, C., Nagar, S., Lal, R., & Negi, R. K. (2018). Fish Gut Microbiome: Current Approaches and Future Perspectives. *Indian Journal of Microbiology*, *58*(4), 397-414. https://doi.org/10.1007/s12088-018-0760-y

[2] Li, X., Ringø, E., Hoseinifar, S. H., Lauzon, H. L., Birkbeck, H., & Yang, D. (2019). The adherence and colonization of microorganisms in fish gastrointestinal tract. *Reviews in Aquaculture*, *11*(3), 603-618.

[3] Ringø, E., Salinas, I., Olsen, R. E., Nyhaug, A., Myklebust, R., & Mayhew, T. M. (2007). Histological changes in intestine of Atlantic salmon (*Salmo salar* L.) following in vitro exposure to pathogenic and probiotic bacterial strains. *Cell and tissue research*, *328*(1), 109-116.

[4] Picchietti, S., Fausto, A. M., Randelli, E., Carnevali, O., Taddei, A. R., Buonocore, F., ... & Abelli, L. (2009). Early treatment with *Lactobacillus delbrueckii* strain induces an increase in intestinal T-cells and granulocytes and modulates immune-related genes of larval *Dicentrarchus labrax* (L.). *Fish & shellfish immunology*, *26*(3), 368-376.

[5] He, S., Zhang, Y., Xu, L., Yang, Y., Marubashi, T., Zhou, Z., & Yao, B. (2013). Effects of dietary *Bacillus subtilis* C-3102 on the production, intestinal cytokine expression and autochthonous bacteria of hybrid tilapia *Oreochromis niloticus♀× Oreochromis aureus♂*. *Aquaculture*, *412*, 125-130.

[6] Newaj-Fyzul, A., Adesiyun, A. A., Mutani, A., Ramsubhag, A., Brunt, J., & Austin, B. (2007). *Bacillus subtilis* AB1 controls Aeromonas infection in rainbow trout (*Oncorhynchus mykiss*, Walbaum). *Journal of applied microbiology*, *103*(5), 1699-1706.

[7] Balcázar, J. L., Vendrell, D., De Blas, I., Ruiz-Zarzuela, I., Gironés, O., & Muzquiz, J. L. (2006). Immune modulation by probiotic strains: quantification of phagocytosis of *Aeromonas salmonicida* by leukocytes isolated from gut of rainbow trout (*Oncorhynchus mykiss*) using a radiolabelling assay. *Comparative immunology, microbiology and infectious diseases*, *29*(5-6), 335-343.

[8] Lazado, C. C., & Caipang, C. M. A. (2014). Mucosal immunity and probiotics in fish. *Fish & shellfish immunology*, *39*(1), 78-89.

[9] Robertson, P. A. W., O'Dowd, C., Burrells, C., Williams, P., & Austin, B. (2000). Use of Carnobacterium sp. as a probiotic for Atlantic salmon (*Salmo salar* L.) and rainbow trout (*Oncorhynchus mykiss*, Walbaum). *Aquaculture*, *185*(3-4), 235-243.

[10] Ray, A. K., Ghosh, K., & Ringø, E. J. A. N. (2012). Enzyme-producing bacteria isolated from fish gut: a review. *Aquaculture Nutrition*, *18*(5), 465-492.

[11] Abid, A., Davies, S. J., Waines, P., Emery, M., Castex, M., Gioacchini, G., ... & Merrifield, D. L. (2013). Dietary synbiotic application modulates Atlantic salmon (*Salmo salar*) intestinal microbial communities and intestinal immunity. *Fish & shellfish immunology*, *35*(6), 1948-1956.

[12] Aftabgard, M., Salarzadeh, A., Mohseni, M., Shabanipour, A. H. B., & Zorriehzahra, M. E. J. (2019). The combined efficiency of dietary isomaltooligosaccharides and Bacillus spp. on the growth, hemato-serological, and intestinal microbiota indices of caspian brown trout (*Salmo trutta* caspius Kessler, 1877). *Probiotics and antimicrobial proteins*, *11*(1), 198-206.

[13] FAO Fisheries & Aquaculture - Cultured Aquatic Species Information Programme - *Salmo salar* (Linnaeus, 1758)

[14] Uren Webster, T. M., Consuegra, S., Hitchings, M., & Garcia de Leaniz, C. (2018). Interpopulation variation in the Atlantic salmon microbiome reflects environmental and genetic diversity. *Applied and Environmental Microbiology*, *84*(16), e00691-18.

[15] Uren Webster, T. M., Rodriguez-Barreto, D., Castaldo, G., Gough, P., Consuegra, S., & Garcia de Leaniz, C. (2020). Environmental plasticity and colonisation history in the Atlantic salmon microbiome: a translocation experiment. *Molecular Ecology*, *29*(5), 886-898.

[16] Llewellyn, M. S., Leadbeater, S., Garcia, C., Sylvain, F. E., Custodio, M., Ang, K. P., ... & Derome, N. (2017). Parasitism perturbs the mucosal microbiome of Atlantic Salmon. *Scientific reports*, *7*(1), 1-10.

[17] Bozzi, D., Rasmussen, J. A., Carøe, C., Sveier, H., Nordøy, K., Gilbert, M. T. P., & Limborg, M. T. (2021). Salmon gut microbiota correlates with disease infection status: potential for monitoring health in farmed animals. *Animal microbiome*, *3*(1), 1-17.

[18] Llewellyn, M. S., McGinnity, P., Dionne, M., Letourneau, J., Thonier, F., Carvalho, G. R., ... & Derome, N. (2016). The biogeography of the Atlantic salmon (*Salmo salar*) gut microbiome. *The ISME journal*, *10*(5), 1280-1284.

[19] Olsen, R. E., Sundell, K., Hansen, T., Hemre, G. I., Myklebust, R., Mayhew, T. M., & Ringø, E. (2002). Acute stress alters the intestinal lining of Atlantic salmon, *Salmo salar* L.: An electron microscopical study. *Fish physiology and biochemistry*, *26*(3), 211-221.

[20] Gajardo, K., Rodiles, A., Kortner, T. M., Krogdahl, Å., Bakke, A. M., Merrifield, D. L., & Sørum, H. (2016). A high-resolution map of the gut microbiota in Atlantic salmon (*Salmo salar*): a basis for comparative gut microbial research. *Scientific reports*, *6*(1), 1-10.

[21] Catalán, N., Villasante, A., Wacyk, J., Ramírez, C., & Romero, J. (2018). Fermented soybean meal increases lactic acid bacteria in gut microbiota of Atlantic salmon (*Salmo salar*). *Probiotics and antimicrobial proteins*, *10*(3), 566-576.

[22] Dehler, C. E., Secombes, C. J., & Martin, S. A. (2017). Environmental and physiological factors shape the gut microbiota of Atlantic salmon parr (*Salmo salar* L.). *Aquaculture*, *467*, 149-157.

[23] Ren, P., Xu, L., Yang, Y., He, S., Liu, W., Ringø, E., & Zhou, Z. (2013). Lactobacillus planarum subsp. plantarum JCM 1149 vs. *Aeromonas hydrophila* NJ-1 in the anterior intestine and posterior intestine of hybrid tilapia *Oreochromis niloticus*♀× *Oreochromis aureus*♂: An ex vivo study. *Fish & shellfish immunology*, *35*(1), 146-153.

[24] Rinkinen, M., Westermarck, E., Salminen, S., & Ouwehand, A. C. (2003). Absence of host specificity for in vitro adhesion of probiotic lactic acid bacteria to intestinal mucus. *Veterinary microbiology*, *97*(1-2), 55-61.

[25] Balcázar, J. L., Vendrell, D., De Blas, I., Ruiz-Zarzuela, I., & Múzquiz, J. L. (2009). Effect of *Lactococcus lactis* CLFP 100 and *Leuconostoc mesenteroides* CLFP 196 on *Aeromonas salmonicida* infection in brown trout (*Salmo trutta*). *Journal of molecular microbiology and biotechnology*, *17*(3), 153-157.

[26] Asgarifar, S. A., Hedayatifard, M., & Khara, H. (2015). Improvement of Growth properties, Survival, hematological-humoral immunity factors and intestinal bacterial density of brown trout (*Salmo trutta* caspius Kessler, 1870) affected by different

levels of Beta-Plus as probiotic. *Aquatics Physiology and Biotechnology*, *3*(3), 1-28.

[27] Ringø, E., Salinas, I., Olsen, R. E., Nyhaug, A., Myklebust, R., & Mayhew, T. M. (2007). Histological changes in intestine of Atlantic salmon (*Salmo salar* L.) following in vitro exposure to pathogenic and probiotic bacterial strains. *Cell and tissue research*, *328*(1), 109-116.

[28] Lashtoaghaei, G., Khara, H., & Sayyad, B. M. (2020). The effect of probiotic Bactocell (*Pediococcus acidilactici*) and iron on growth in Caspian Sea Salmon (*Salmo trutta* caspius).

[29] Hosseini, A., Oraji, H., Yegane, S., & Shahabi, H. (2014). The effect of probiotic *Pediococcus acidilactici* on growth performance, blood and some serum parameters in Caspian salmon (*Salmo trutta* caspius).

[30] Oraji, H., Yegane, S., & Shahabi, H. (2014). The effect of probiotic bactocell on growth performance, blood parameters and some serum parameters in Caspian salmon (Salmo caspius). *ISFJ*, *23*(2), 35-44.

[31] Jami, M. J., Kenari, A. A., Paknejad, H., & Mohseni, M. (2019). Effects of dietary b-glucan, mannan oligosaccharide, *Lactobacillus plantarum* and their combinations on growth performance, immunity and immune related gene expression of Caspian trout, *Salmo trutta* caspius (Kessler, 1877). *Fish & shellfish immunology*, *91*, 202-208.

[32] Kristiansen, M., & Ringø, E. (2011). Evaluation of prebiotic and probiotic effects on the intestinal gut microbiota and histology of Atlantic salmon (*Salmo salar* L.).

[33] Jaramillo-Torres, A., Rawling, M. D., Rodiles, A., Mikalsen, H. E., Johansen, L. H., Tinsley, J., ... & Merrifield, D. L. (2019). Influence of dietary supplementation of probiotic *Pediococcus acidilactici* MA18/5M during the transition from freshwater to seawater on intestinal health and microbiota of Atlantic salmon (*Salmo salar* L.). *Frontiers in microbiology*, *10*, 2243.

[34] Salinas, I., Myklebust, R., Esteban, M. A., Olsen, R. E., Meseguer, J., & Ringø, E. (2008). In vitro studies of *Lactobacillus delbrueckii* subsp. lactis in Atlantic salmon (*Salmo salar* L.) foregut: tissue responses and evidence of protection against *Aeromonas salmonicida* subsp. salmonicida epithelial damage. *Veterinary microbiology*, *128*(1-2), 167-177.

[35] Leisner, J. J., Laursen, B. G., Prévost, H., Drider, D., & Dalgaard, P. (2007). Carnobacterium: positive and negative effects in the environment and in foods. *FEMS microbiology reviews*, *31*(5), 592-613. https://doi.org/10.1111/j.1574-6976.2007.00080.x

[36] Roh, H., Kim, B. S., Lee, M. K., Park, C. I., & Kim, D. H. (2020). Genome-wide comparison of *Carnobacterium maltaromaticum* derived from diseased fish harbouring important virulence-related genes. *Journal of Fish Diseases*, *43*(9), 1029-1037.

[37] Garcés, M. E., Olivera, N. L., Fernández, M., Riva Rossi, C., & Sequeiros, C. (2020). Antimicrobial activity of bacteriocin-producing Carnobacterium spp. isolated from healthy Patagonian trout and their potential for use in aquaculture. *Aquaculture Research*, *51*(11), 4602-4612.

[38] Nguyen, C. D., Amoroso, G., Ventura, T., & Elizur, A. (2020). Assessing the Pyloric Caeca and Distal Gut Microbiota Correlation with Flesh Color in Atlantic Salmon (*Salmo salar* L., 1758). *Microorganisms*, *8*(8), 1244.

[39] Skrede, G., Storebakken, T., Skrede, A., Sahlstrøm, S., Sørensen, M., Shearer, K. D., & Slinde, E. (2002). Lactic acid fermentation of wheat and barley whole meal flours improves digestibility of nutrients and energy in Atlantic salmon (*Salmo salar* L.) diets. *Aquaculture*, *210*(1-4), 305-321.

[40] Porto, M. C. W., Kuniyoshi, T. M., Azevedo, P. O. S., Vitolo, M., & Oliveira, R. S. (2017). Pediococcus spp.: An important genus of lactic acid bacteria and pediocin producers. *Biotechnology Advances*, *35*(3), 361-374.

[41] Giri, S. S., Sukumaran, V., Sen, S. S., & Jena, P. K. (2014). Effects of dietary supplementation of potential probiotic *Bacillus subtilis* VSG 1 singularly or in combination with *Lactobacillus plantarum* VSG 3 or/and *Pseudomonas aeruginosa* VSG 2 on the growth, immunity and disease resistance of *Labeo rohita*. *Aquaculture Nutrition*, *20*(2), 163-171.

[42] Liu, C. H., Chiu, C. H., Wang, S. W., & Cheng, W. (2012). Dietary administration of the probiotic, *Bacillus subtilis* E20, enhances the growth, innate immune responses, and disease resistance of the grouper, *Epinephelus coioides*. *Fish & shellfish immunology*, *33*(4), 699-706.

[43] Aftabgard, M., Salarzadeh, A., & Mohseni, M. (2019). The effects of a synbiotic mixture of galacto-oligosaccharides and Bacillus strains in Caspian Salmon, *Salmo trutta* caspius Fingerlings. *Probiotics and antimicrobial proteins*, *11*(4), 1300-1308.

[44] Can, E., Kurtoğlu, İ. Z., Benzer, F., Erişir, M., Kocabaş, M., Kızak, V., ... & Çelik, H. T. (2012). The effects of different dosage of kefir with different durations on growth performances and antioxidant system in the blood and liver tissues of Çoruh trout (Salmo coruhensis). *Turkish Journal of Fisheries and Aquatic Sciences*, *12*(2), 277-283.

[45] Can, E., Kutluyer, F., Sonay, F. D., & Köse, Ö. (2012). The use of kefir as potential probiotic in Çoruh trout (Salmo coruhensis): Effects on growth performance and immunoglobulin (IgM) levels. *African Journal of Biotechnology*, *11*(30), 7775-7780.

[46] Can, S. S., Kutluyer, F., Can, E., KAYIŞ, Ş., Delihasan Sonay, F., Köse, Ö., ... & Kayim, M. (2014). Effect of dietary kefir on the digestive and liver enzymes activities and glucose level of Coruh trout, Salmo coruhensis (Actinopterygii: Salmoniformes: Salmonidae). *Acta Ichthyologica et Piscatoria*, *44*(2).

[47] Sotoudeh, E., Amiri Moghaddam, J., Shahhosseini, G., & Aramli, M. S. (2016). Effect of Dietary Gamma-irradiated and Fermented Soybean Meal on the Growth Performance, Body Composition, and Digestive Enzymes Activity of Caspian Brown Trout, *Salmo trutta* caspius, Juvenile. *Journal of the World Aquaculture Society*, *47*(6), 830-842.

[48] Voorhees, J. M., Barnes, M. E., Chipps, S. R., & Brown, M. L. (2018). Rearing performance of juvenile brown trout (*Salmo trutta*) subjected to exercise and dietary bioprocessed soybean meal. *Open Journal of Animal Sciences*, *8*(3), 303-328.

[49] Moghaddam, A., & Sadegh, M. (2016). Effect of Dietary Gammaâ irradiated and Fermented Soybean Meal on the Growth Performance, Body Composition, and Digestive Enzymes Activity of Caspian Brown Trout, *Salmo trutta caspius*, Juvenile. *Journal of the World Aquaculture Society*.

[50] Kumazawa, T., Nishimura, A., Asai, N., & Adachi, T. (2018). Isolation of immune-regulatory *Tetragenococcus halophilus* from miso. *PloS one*, *13*(12), e0208821.

[51] Wang, C., Liu, Y., Sun, G., Li, X., & Liu, Z. (2019). Growth, immune

response, antioxidant capability, and disease resistance of juvenile Atlantic salmon (*Salmo salar* L.) fed *Bacillus velezensis* V4 and *Rhodotorula mucilaginosa* compound. *Aquaculture*, *500*, 65-74.

[52] Salinas, I., Cuesta, A., Esteban, M.A., Meseguer, J., 2005. Dietary administration of *Lactobacillus delbrueckii* and *Bacillus subtilis*, single or combined, on gilthead seabream cellular innate immune responses. *Fish & Shellfish Immunology,* 19, 67-77

[53] Kiron, V., Kulkarni, A., Dahle, D., Lokesh, J., & Kitani, Y. (2015). A microbial feed additive abates intestinal inflammation in Atlantic salmon. *Frontiers in immunology*, *6*, 409.

[54] Allwood, J. G., Wakeling, L. T., & Bean, D. C. (2021). Fermentation and the microbial community of Japanese koji and miso: A review. *Journal of Food Science*.

Section 2

Issues in Salmon Aquaculture Industry

Chapter 4

Perspectives on Salmon Aquaculture: Current Status, Challenges and Genetic Improvement for Future Growth

James E. Barasa, Purity Nasimiyu Mukhongo and Cynthia Chepkemoi Ngetich

Abstract

With an estimated global value of US$15.6 billion, farmed salmonids represent a precious food resource, which is also the fastest increasing food producing industry with annual growth of 7% in production. A total average of 3,594,000 metric tonnes was produced in 2020, behind Chinese and Indian carps, tilapias and catfishes. Lead producers of farmed salmonids are Norway, Chile, Faroe, Canada and Scotland, stimulated by increasing global demand and market. However, over the last 2 years, production has been declining, occasioned by effects of diseases as well as rising feed costs. Over the last year, production has declined sharply due to effects of covid-19. This chapter reviews the species in culture, systems of culture, environmental footprints of salmon culture, and market trends in salmon culture. Burden of diseases, especially Infectious pancreatic Necrosis, Infectious salmon anemia and furunculosis, as well as high cost of feed formulation, key challenges curtailing growth of the salmon production industry, are discussed. A review is made of the international salmon genome sequencing effort, selective breeding for disease resistance, and the use of genomics to mitigate challenges of diseases that stifle higher production of salmonids globally.

Keywords: salmon, smolts, salmon genome, fish meal, parr, anadromous

1. Introduction

Salmonids constitute a large group of teleost fishes thriving in the cold-water fisheries and aquaculture. Salmonids belong to the family Salmonidae, comprising 11 genera, including the salmon, trout, charr, ciscos, grayling, hucho and the freshwater whitefish [1]. The sub-family Salmoninae groups three well known genera: *Onchorynchus*: rainbow trout, cutthroat trout, Pacific salmon, all with native ranges in the North Pacific Ocean. The genus *Salmo* groups the Atlantic salmon, Atlantic trout and the brown trout, all with native ranges in the North Atlantic Ocean. *Salvelinus* comprises the charr, with native ranges in the Pacific shores. The Pacific salmon has 5 species: chinook (*Onchorynchus tshawytsha*), chum (*O. keta*), coho (*O. kisutch*), casu (*O. masu*), pink (*O. gorbuscha*) and sockeye (*O. nerka*) [2]. Pacific salmon are basically anadromous (migrate to sea water after spending early life in streams and rivers), semelparous (reproduce only once in a life time) and exhibit accurate homing. Atlantic salmon which inhabits the eastern coast of North America, are homing and iteroparous (reproduce more than once in a life time) [3].

Salmonids are the third largest farmed fin fish crop, behind Chinese and Indian carps and tilapias, with a total annual production of 3,594,000 metric tonnes [4]. They however form the lead farmed carnivorous fin fish globally. Production of salmon (Atlantic and Pacific salmon) forms the fastest growing food producing industry in the world, with annual growth of 7%. Atlantic salmon, *Salmo salar*, is iconic, high value, widely traded global fish product, and natural stocks are often threatened by overexploitation and habitat degradation [5]. It contributes substantially to food, economic and employment security in many countries, especially Norway, Chile, Canada and the United Kingdom [5], which are lead producers of the species (**Table 1** and **Figure 1**).

Significant development in the farming of *S. salar* is recorded in temperate coastal regions of countries such as Norway, Canada and Scotland [6], with Chile being among the top producers. A total of 30,000 direct and over 14,500 indirect jobs are provided by the salmon industry in Chile [7], underscoring the importance of salmon industry in the country, which is also the second biggest producer of farmed salmonids globally, with annual production averages of 700,000 tonnes [8]. Farmed salmonids account for over 73% of aquaculture production in Chile and became the second largest contributor to the Chilean economy [9], with the three most intensively farmed salmon species being *S. salar*, *O. mykiss* and *O. kisutch* [10]. In Scotland, salmon farming takes place on the west coast and islands of Scotland and approximately 95% of the aquaculture industry is dominated by *S. salar*, making it the third largest producer after Chile and Norway [7]. These countries are located within certain southern hemispheres that are at a constant temperature of around 0–20°C. Salmon farming ideally requires temperature of 13°C [11], or below. But the fish's appetite for food reduces at very low temperatures, and can therefore affect growth rates. Typically, juvenile fish less than 250 g (raised in freshwater) are released in to pens or cages in the ocean, where they are grown to market size of 2–8 kg a piece, within a grow-out period of 16–24 months. Before reaching 250 g for release to pens or cages, the early forms grow in wide areas of freshwater farms and hatcheries across eastern and southern Chile [12]. Thereafter, they are released for fattening in the marine environments in the southern most Patagonian fjords [13]. These areas are endowed with ample water flow in current, protected naturally by fjords and archipelagos.

With average growth in annual production of about 9%, salmonids represent the fastest growing food production system globally over the years, highlighting the important role of the fish in food and nutrition security, as well as livelihood and income generation. Salmon is rich in protein, omega-3 fatty acids, minerals and vitamins. In this respect, the Atlantic salmon is iconic in value, distribution and conservation status. With an increasing demand globally, consumption of salmon is currently 3 times its quantity for 1980, and contributes 70% of the market for salmonids. Apart from its high global demand, the high visibility of salmon on the market is due to the high level of industrialization and low-level risk associated with its culture. Contrary to its status of a luxury commodity in the 1980s, it is a major food item in the USA, Europe and Japan, with high prices of about US$11.9 in USA and US$ 7.3 per kg in Europe. High demand is also driven by lucrative emerging markets in China, Russia, and Brazil [14]. Farming *S. salar* is also much more efficient, about 8 times more efficient than beef production.

Species	2017		2018		2019		2020	
	Tonnage	% growth	Tonnage	% growth	Tonnage	% growth	Tonnage	% growth
Chinese carps	19,131		19,469		20,090		21,747	
Tilapias	5381		6276		6513		6800	
Catfishes	4553	7.2	4879	2.5	5003	3.8	5193	
Atlantic salmon	2,290,000	5.8	2,423,000	7.3	2,599,000	3.5	2,689,000	
Coho salmon	171	9.6	187	8.6	203	2.5	208	
Large rainbow trout	261	1.6	265	13.1	300	0.7	302	
Small rainbow trout	582	2.5	596	2.4	610	2.0	623	
Pangasius	1249	13.8	1422	3.3	1468	2.6	1506	
Sea bass and sea bream	403	4.0	419	−1.7	412		387	

Table 1.
Global production of lead farmed fish species 2017–2020 (000 metric tonnes).

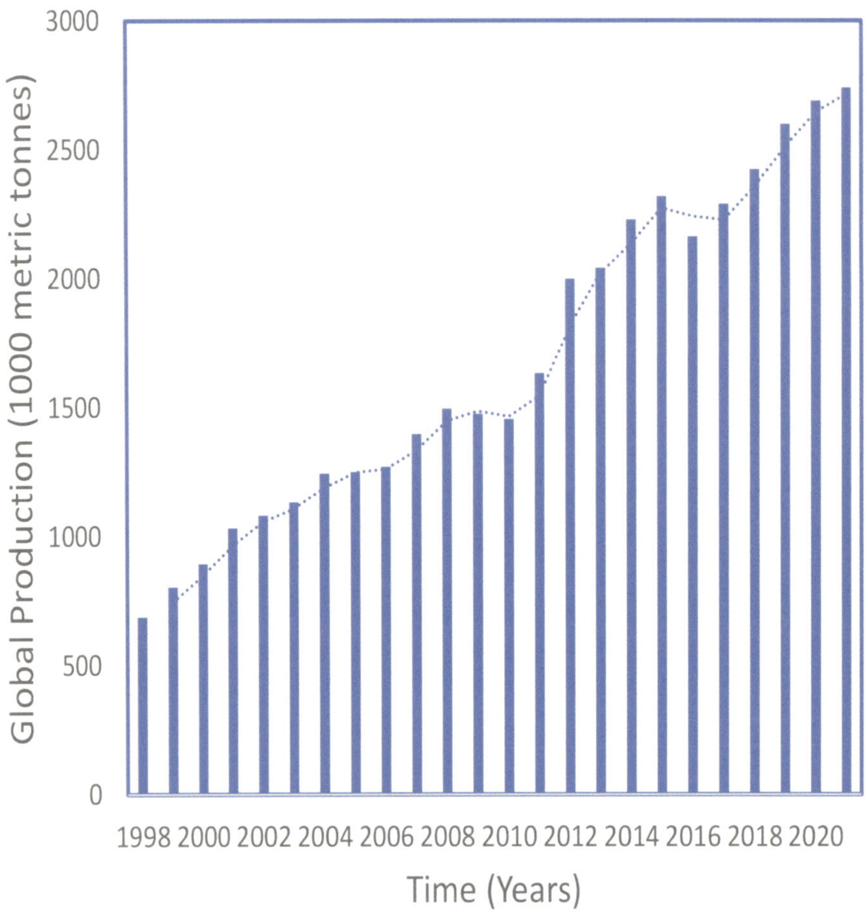

Figure 1.
Global production of farmed Salmo salar *(1998–2021), in metric tonnes. A steady increase in production is recorded annually over the years, demonstrating the importance of the species as food and source of income in main global producing countries.*

1.1 Anadromy in Atlantic salmon

Most salmonid species are anadromous. Hence, they spawn in freshwaters (streams, lakes and rivers), where the young ones spend 1–3 years before juvenile stages migrate to the sea for feeding and fattening. The ability to switch lifestyle from freshwater to sea water is called smoltification, a process controlled by temperature and photoperiod. As the fish mature in the sea, they begin to return to their point of release or spawn, in a process called homing, for spawning. Most salmons are semelparous, i.e. they breed only once in their life time and then die. Death is mainly due to exhaustion from the long distances covered during homing, and the excessive energy spent during spawning. A few salmons are iteroparous, i.e. spawn severally in their lifetime. This is because they are able to migrate back to the sea, after spawning to continue feeding and rebuilding their reproductive capacity. However, some species, such as rainbow trout complete their life cycles in freshwater. Anadromous salmon shows fidelity to the freshwater site at which they were spawned, or released, and therefore when they reach sexual maturity and are about to start breeding, they migrate back to these sites for spawning. The return of mature salmon to their natal streams from the ocean is called homing, a complex process in which majority of the fish return to their actual natal streams, while a few veer off to different streams.

Suitable environmental conditions for growth of salmon include: low water temperatures of 8–16°C, clean and well aerated waters and well protected fjords, free from storms and other environmental upheavals [14]. These low temperatures reduce stress to the fish during summer, and reduce the growth rate in winter, conditions suitable for minimizing disease incidences among salmonids. Typically, these conditions are found in Norway, Chile, the North Atlantic and North Pacific coasts, as well as coastlines of Tasmania and New Zealand. These are countries of higher latitudinal ranges, often temperate regions. Although it is generally regarded that fish production increases with reducing latitudes, especially for warm water species [15], production of cold-water species nevertheless positively increases with increasing latitudes [15]. Although for warm water species, the effect of temperature on fish production in a fishery is generally boosted by the fertility of the waters [16], temperature is probably the main factor driving productivity of salmonids [15], given that most salmonid species are generally farmed in sea ranches or raceways on fish farms, systems that require clean, well aerated waters.

As high value species of global demand in the developed countries, salmonids are usually cultured in intensive systems, characterized by high fish densities, low water flow, and high concentrations of dissolved oxygen. Removal of carbon dioxide, solids and excretion end-products, such as ammonia and nitrites, are generally prioritized, in order to improve growth and health of the fish. Generally, *S. salar* has a low tolerance to a dissolved oxygen deficit, sensitive to increased concentrations of carbon dioxide, un-ionized ammonia and nitrites in freshwater.

2. Main aquaculture systems used in salmon production

2.1 Flow through systems (FTS)

Flow through systems, also called raceways or semi-closed culture systems, are culture units in which water flows continuously, making a single pass through the unit before being discharged. Raceways are mainly concrete, but some are earthen, lined with waterproof materials, yet some are fabricated from wood, fiber glass, metal, plastic and other materials, depending on the resources of the farmer. They are majorly designed for highly intensive culture and especially suitable for fish species that need constantly flowing clean water e.g. juvenile salmon and in the production of smolts. As high value species therefore, almost 80% of *S. salar* smolts globally are produced in flow through systems before being stocked in sea cages. Egg-larvae (parr) are supplied with fresh water from local sources such as rivers, lakes, ground water or natural springs at hatcheries and fish farms. Good flow rates and velocity of water is essential to the health of the stock under culture, and to flush wastes from the system. Water quality is maintained by treatment and manipulation of the flow rate of the water. The quality is then enhanced by injection of oxygen using air blowers, in order to minimize the water flow rate to 0.6 L/kg/min. Sophisticated farms heat the water to a certain temperature and manipulate light intensity [6]. Appropriate stocking density of the fish in FTS is dependent on water quality, management skills and general husbandry practices put in place, as well as the biology of the species, including the ability of the fish to tolerate crowding. FTS are capable of supporting a high number of smolts yearly, with averages of 900 million smolts in Norway alone produced under FTS [17]. Fish reared in these systems can however be highly susceptible to diseases due to stress caused by overcrowding. Raceway systems can be earthen or concrete based, majority are constructed from concrete or cement blocks.

2.2 Recirculation system (RAS)

Recirculation system was developed in the 1970s, to reduce the required amount of water and resultant waste produced from traditional flow through system. The system is highly controlled, and therefore requires substantial skills and input. In order to reduce the demand for large amounts of water, the system involves recirculation of water, which also reduces water wastage. Since water is recirculated in the system, there is enhanced biosecurity on salmon fish farms and hatcheries, and this prevents or minimizes escapee fish to the natural environment. Prevention of escapees from farms or hatcheries comes with several benefits, such as preservation of the purity of local natural populations of salmon, reduced incidences of disease and parasites to the natural populations, as well as to those within farms [18]. Wastes from fish are controlled and easily collected, which reduces pollution of the environment and the collected waste is easily aggregated for subsequent use for other purposes on the farm. Additionally, the environment for fish growth is optimized, with control of water temperature, water quality, feeds, and these maximize growth rates of the fish. Due to its ability to minimize impacts to the environment, RAS is easily and locally sited to markets, which therefore reduces transportation costs and carbon footprints, while simultaneously improving traceability and freshness of the product, and profitability of the enterprise as well.

Improved RAS systems comprise of two portions, with one part of the tank dedicated to fish rearing and trapping of particles and draining of sludge [19]. The other part is the water treatment system, composed of an additional solid removal system, submerged biofilter and an airlift for water circulation and gas exchange [19]. This therefore allows addition of oxygen and removal of carbon dioxide and ammonia gases from the water. Additional mechanical filters aid the removal of particles that would not settle. Generally, the efficiency of the RAS is enhanced, in order to reduce the amount of energy required to produce a kg of fish, while maximizing the stocking density of the fish (typically 61–122 kg/m^3, but in some cases, may exceed 545 kg/m^3). The system is therefore successfully applied to rear smolts in many salmon producing countries [20]. In Norway, a total of 12–20 million smolt per year are produced under RAS [21]. Averagely, 350,000 MT of salmon are produced annually under RAS [22]. High technological complexities that necessitate high costs of production and highly skilled and competent human resource is the key challenge facing salmon farmers that operate RAS.

2.3 Cage culture

Cages or pens are natural or semi-sheltered bay where the shoreline forms all but one side of the enclosure. Cages or pens are made from bamboo, wooden poles or stakes driven into the substrate, the mesh size is typically small enough to retain the cultured fish but large enough to allow entry and exit of small fish and food organisms. Its management is less complex than land-based systems, make use of existing water bodies which gives local non-land owners access to fish farming. This type of system makes the majority of the salmon grow out particularly for seawater operations, and is appealing to most farmers, for incurring the lowest production and operation costs of all the production systems.

Cages are movable and float off the bottom, range from about 1 m^2 to over 1000 m^2 in surface area, with a depth of about 20–50 m, and a maximum circumference of 157 m. The stocking density limits for post smolt *S. salar* in commercial scale culture averages 75 kg/m^3 [23]. Average production volumes almost doubled in Norway, increasing from 37 to 67 million m^3 from the year 2005–2009, mainly due to better quality of water, better food organisms and reduced impact of storms.

Escaped fish, predation by seals and climate change are the main challenges facing cage culture of salmon.

3. Marketing of salmon

As an iconic group of fishes, salmon is very rich in high quality proteins, and long chain omega-3 fatty acids, which reduce the risk of cardiovascular disease and other health issues. It is a good source of minerals (iodine and selenium), vitamins (D and B12) and macronutrients. Due to this important nutritional composition, salmon is a globally traded product, especially in the developed countries, where purchasing power is also high.

3.1 Processing of salmon

In order to increase safety of the product, preserve high quality characteristics, extend shelf life and enhance economic returns to the producer, salmon is processed in different forms [24]. About 47% of the EU market supply of salmon is filleted, while 12% is of whole fish form, which is also the most preferred since they are fresh and preserved through chilling and freezing. A total of 28% of the supply is smoked salmon, while 13% constitute other value-added products [25]. Smoked salmon is the most expensive, sold at €90 a kilo, while fillets cost €14 a kilo [25]. Processing plants are required to ensure that the weight, color, size, shape and packaging of the final product are of the standard desired by the final consumer. This requires well trained and skilled workers to assure product quality. In this regard, and in an effort to maintain the highest standards of safety and hygiene of this globally traded fish product, processing facilities are often certified by US and EU authorities for them to qualify to supply export markets (**Figure 2**). Some of the requirements for this certification is the maintenance of solid cold chains, international standards of germ

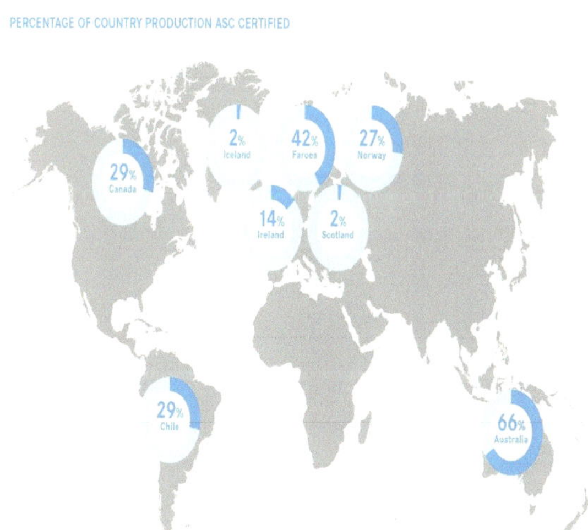

Figure 2.
Percentage of salmon producing companies in each of the main global salmon producing countries that are certified by Aquaculture Stewardship Council (ASC). Some of the criteria used by the ASC for certification of production value chain includes: the amount of fish meal used in formulating fish feed and fish oil for the farmed salmon, the amount of chemicals and drugs used in control of parasites and diseases, and biosecurity or the level of control put in place on fish farms to limit escape of farmed fish to the natural environment, lethal incidents involving marine mammals, antibiotic use and viral disease mortality. Fish from certified farms should be more attractive to export markets.

control, i.e. the Hazard Analysis of Critical Control Points (HACCP) certification and efficient systems of waste management. As happens with other fish products, salmon processors also undertake value addition, to increase the shelf-life and value as well as expand the market [26]. The main value-added products of salmon include fillets, salmon bread, sushi, and smoked salmon [26]. Apart from improved purchasing power and awareness of nutritional benefits of consumption of salmon, this hygienic standards in processing the product and value addition have seen increased consumption of the product (**Figure 3**). A total consumption of farmed Atlantic salmon of 2.4 million tonnes was estimated in 2020 [4], which, when combined with those from capture fisheries rises to 3.2 million tonnes.

3.2 Packaging of salmon

Packaging is crucial for providing useful information to the consumer, such as product identity, origin, how to use and store, nutritional information among others. Well packaged fish products enhance efficient mechanized handling, distribution and marketing. Rigid materials like cans, glass container jars, plastic bags, pouches, film, sheets, jars and boxes are commonly used in packing salmon [27]. Fresh fish are usually loaded in plastic boxes that are hygienic, light and strong. The boxes are insulated to maintain the temperature of iced fish, while also allowing drainage of any melted liquid from the fish [27]. Frozen fish is commonly packed in interlocking, printed, polycoated and corrugated fiberboard cartons and expanded polystyrene and corrugated polypropylene boxes, sealed with polypropylene or metal tape. This type of boxes are also used for freezing wet fish, storing wrapped or unwrapped frozen fish [27]. Fresh, chilled or frozen fish are packed using Styrofoam, polyvinylidene chloride or polystyrene trays wrapped with cling film made from either polythene or polypropylene. Although this type of packaging can

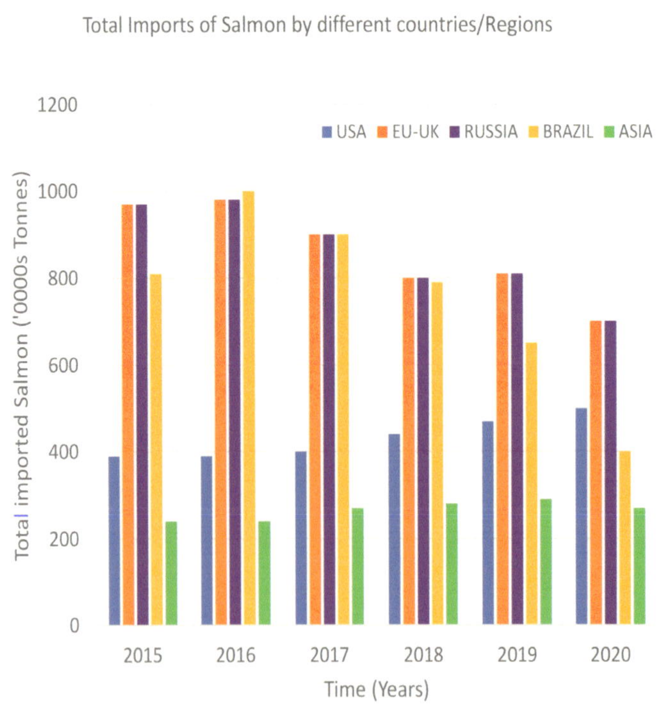

Figure 3.
Total imports of salmon by major consuming countries or regions from 2015 to 2020. USA is the United States of America, EU-UK is the European Union and the United Kingdom.

be attractive to customers, they cannot protect the fish from mechanical damage, loss of moisture and aroma or even contamination from microorganisms and odor from other products [27].

3.3 Freight packaging

Fresh, frozen or live salmon for airfreight is packaged in containers made from metal, fiberglass and expanded polystyrene. Such a container is insulated, easy to handle, heavy to give physical protection to the products and watertight to protect against contamination [27].

The main importing countries or regions for salmon products include the USA, EU-UK, Russia, Brazil and Asia. Although imports or consumption of salmon has been decreasing in the EU-UK, Russia and Brazil since 2016, consumption of salmon products has been on the increase in the USA and Asia (**Figure 3**), causing an increase in imports. The decline in Russian imports is occasioned by an embargo on salmon imports from Norway following the EU's trade sanctions against Russia due to the conflict in Ukrain.

4. Main challenges in salmon aquaculture

4.1 Incidences of diseases

The main diseases in farmed salmons include infectious salmon anemia (ISA), characterized by pale gills and fish that swim close to the water surface while gulping for air. Some cases are asymptomatic, but the fish die suddenly. ISA was first reported in Norwegian salmon farms in 1984, from where it spread to other big producers of salmon, causing huge losses of up to €100 million [28, 29]. ISA is caused by a virus, the Infectious salmon anemia virus, one of the most devastating diseases of marine farmed *S. salar*, and mainly attacks the grow out stages of the fish. In Chile, the first outbreak was in June 2007, with the ISAV HPR7b variant in circulation [30]. The impact of the outbreak was devastating, and was partly responsible for the brief decline in global production of salmon in subsequent years (**Figure 4**). ISA outbreaks come with high mortality of fish, huge loses to farmers and severe restriction to production in surrounding areas. A large number of risk factors are known to predispose salmon to ISA outbreaks [32]. Presence of the ISAV receptors in the fish, the variant strain (whether virulent or non-virulent) responsible for an outbreak, rate of evolution of the strain from non-virulent to virulent, the rate of viral reproduction and shedding, suboptimal management practices at cage farms, fish stocking and fallowing routines in cages, related disease outbreak events, level of intensification in fish production on the farm, and handling and treatment of fish constitute some risk factors that fuel increased incidences of ISA outbreaks [33–35]. Increased biosecurity, advanced fish husbandry practices, as well as a better understanding of some of the risk factors constitute suitable mitigation measures for ISA outbreaks [32].

Infectious pancreatic necrosis (IPN) is a disease of young salmonids (*Salmo*, *Onchorynchus* and *Salvelinus*), attacking the pancreas and liver parenchyma of the fish. The virus responsible for IPN is an *Aquabinarvirus* of family Birnaviridae, and comprises a bi-segmented double stranded RNA. Severe necrosis of pancreatic and liver cells occurs, which extends to the intestinal mucosae [36]. Post-smolts darken in color, anterior part of the abdomen swells, capillaries around the pectoral fins engorge, the dorsal fin erodes, while the vent swells [36]. It occurs both in freshwater and marine water stages, when the fish is typically of start-feeding stage to about 20 g (after

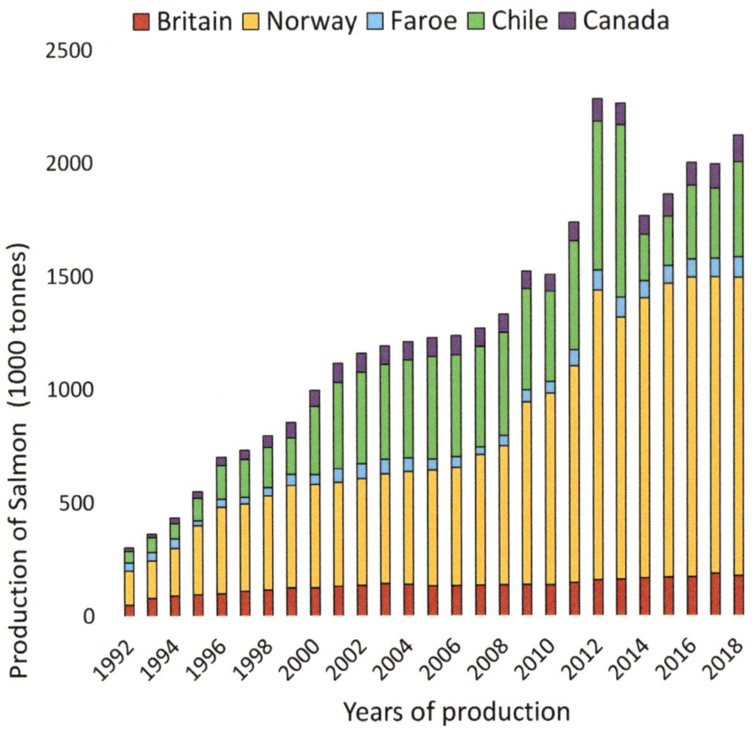

Figure 4.
*Production of Atlantic salmon (*Salmo salar*) in the 5 largest producer countries from 1992 to 2018. Norway remains the lead producer, followed by Chile over the years. Adopted from Iversen et al. [31].*

transfer to the seas in post-smoltification stage). Therefore, the disease attacks fry and post-smolts, and becomes especially severe in the marine environments (post-smolts), causing substantial mortality [37]. Effects of IPNV outbreaks are therefore economic, ecological, and social (welfare), since the salmon that survive the attack often remain asymptomatic carriers of the virus [38]. Economic losses due to IPNV outbreaks in Norway, for instance were estimated at US$ 30 million [39]. Apart from presence in the pancreatic cells, some of the virus cells hide and therefore multiply and persist in the leucocytes of the head kidney. This leads to recurrent outbreaks, which spread quickly across farms in a locality, especially in lead salmon producers like Norway, and Chile where farms are concentrated in a locality. Some of the host defense mechanisms against IPNV include the interferon necrosis (IFN) factor and the anti-viral protein or gene Mx [40], which suppress the persisting viral cells. Following the introduction of IPN resistant strain (IPN-QTL) homozygous in salmon producing countries, mortality now varies based on the susceptible and resistant strains of salmon during the outbreak. Therefore, mortality can vary from 5 to 10% in the resistant strains of salmon, to 70% in the genetically susceptible strains in sea cages. This suggests that considerable gains against diseases in farmed salmon production can be made by a combination of selective breeding of salmon for disease resistance and a suite of both natural and active immune responses against invading pathogens.

Apart from infecting salmonids, furunculosis is also highly pathogenic for other fish species of the wild waters as well as farmed populations. It is caused by a gram-negative rod bacterium, *Aeromonas salmonicida* subsp. *salmonicida*. The bacterium carries an external surface layer, the A-protein surface layer (A-layer), which counters the host defense mechanisms of the fish. This is boosted by a lipopolysaccharide, a protective cell envelope antigen on the surface of the bacterium [41]. As the bacteria grow, they release extracellular products, which cause lesions on the fish. Therefore, symptomatic cases are characterized by fish with lesions that lead to mortality in severe cases [41]. Infected fish are generally lethargic, lack appetite, develop dark skins, show ventral haemorhage at the base of anal, pectoral and pelvic fins, splenomegaly and subcapsular haemorhage occur in the liver [41]. When liquefactive skin lesions and ulcers rupture, more bacteria are released in to the environment, and increase infection of the surrounding fish. Severe outbreaks are reported to cause economic losses in excess of US$100 million in Norway salmon Industry [28]. Control measures during outbreaks include prophylaxis, such as use of vaccines. Drugs (antibiotics such as flumequina) against furunculosis may be administered to the infected fish through diets, while best management practices are recommended to avoid outbreaks. In case of severe outbreaks, movement of smolts may be banned (quarantining farms), and farms that suffer outbreaks are banned from sale of smolts [41]. Common risk factors that induce outbreaks of furunculosis in salmon farms include: migration of fish, water quality, sharing or transfer of staff among salmon farms and hatcheries, breach of quarantine protocols and poor husbandry and hygienic practices of hatcheries and farms [42]. Additionally, algal blooms, increasing temperatures and salinity in wild waters increase the risk of outbreak of furunculosis [43].

4.2 High cost of feeds for salmon production

As a carnivorous fish group, farmed salmonids require high quality feeds (high crude protein content) for fast growth, and to attain appropriate nutritional composition. Typically, feeds for salmon comprise of: 93.4% dry matter, 35.6% crude protein, 33.5% crude lipid, 11.0% carbohydrates and 1.3% phosphorus [44]. However, formulating and maintaining such high-quality diets is not only expensive, but also environmentally challenging, as it requires high amounts of marine fish resources to provide the ingredients for protein and oils, which invariably increases overexploitation of resources (overfishing). In this regard, formulation of suitable diets for farmed salmon requires inclusion of fish meal and fish oil in appropriate quantities, to give the final product the required nutritional quality and composition. Usually, formulated diets for salmon constitute 40–60% fish meal and 20–30% fish oil, sourced mainly from marine anchovies, mackerel, pilchards, herring and blue whiting [45]. These marine fish species are often targeted as sources of fish meal and fish oil for salmon feed production because they provide appropriate nutrients for carnivorous fish species and offer appropriate amounts of polyunsaturated fatty acids (omega 3), in the fillets of the salmon, which is beneficial for human health. Notwithstanding the benefits of using fish meal and oils in salmon diet for human health, the practice not only makes salmon diets expensive, but also increases overfishing of target marine fish species, and so runs contrary to sound principles of conservation of aquatic biodiversity. Since the mid-2000s, the prices of fish meal and fish oil rose between 50 and 130% [46]. Such increases in the costs of key ingredients, coupled with the fact that traditionally, fish feeds form the highest cost of total fish culture enterprises, feeds for farmed salmon production provide a critical challenge in the global culture of salmonids, for their high cost and unsustainability in the long term. Previous studies report an intake of 2.5 kg of marine fish to produce 1 kg of salmon [46]. Globally, 1 kg of salmon feed retails at an average price of NOK 13 (€1).

In order to address this challenge, and increase efficiency and sustainability of farmed salmon production, viability lies in diversifying the sources of protein and oils, especially plant sources, in order to reduce exploitation of marine fish species for fish meal and oils, but still retain high nutritional quality of the diets. In this regard, the composition of formulated feeds for salmon has been changing since

1990, with some of the marine ingredients being replaced by ingredients from plant sources [44]. Both fish meal and fish oil composition in salmon feed formulation have declined, with replacement by plant-based ingredients (**Figure 5**). Studies in to alternative feed resources report suitability of zooplankton, mesopelagic fish, some species of squids, and the Antarctic and North Atlantic krill as viable alternatives to fish meal and fish oils [47], as they equally supply excellent levels of omega-3 poly-unsaturated fatty acids, vitamins, minerals, essential amino acids, carotenoids and nucleotides. The nutritional composition of such alternative diets is enhanced further by feed additives [47], prebiotics and immunostimulants. Another appealing alternative is the use of by-products and by-catch (non-target fish and other aquatic organisms caught during fishing) from fisheries and aquaculture. This targets the utilization of non-edible parts of fish from processing plants, as well as the discards from fishing expeditions. The use of these materials as ingredients in formulating diets for salmon is strictly undertaken in conformity with the regulations in place, such as the EU regulations on the use of animal products, to control and prevent the spread of diseases and bioaccumulation of contaminants and other undesirable substances [45]. As long as the selection of such materials is done properly, taking in to consideration their nutritional composition, they impart useful nutrients to the feeds formulated for farmed salmon, helping achieve cost-effectiveness, sustainability and high quality of diets, without the use of fish meal and oils [45].

Efforts to find viable alternatives to fish meal and fish oils in feeds for salmon have been concentrated on plant products or ingredients, since their availability, nutritional quality and prices can be achieved competitively. This is underscored by the increasing amounts of plant matter used as ingredients in formulation of feeds for farmed salmon, in comparison with fish meal and fish oils, which are on the decline (**Figure 5** and **Table 2**). In this regard, one of the most suitable and promising plant ingredients is the soy beans as the source of protein and oils, for its high

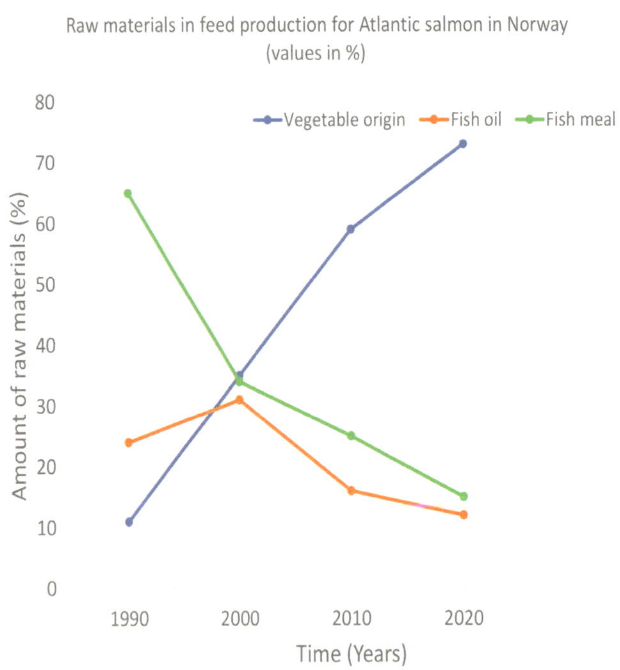

Figure 5.
Trends for raw materials used in feed production for Atlantic salmon in Norway (values in %). Since 1990, vegetable-based ingredients are often used to replace fish meal in feeds for salmon, in order to reduce overexploitation of marine fish species.

Ingredient	% composition
Plant protein sources	40.2
Plant oils	20.1
Carbohydrate sources	10.7
Marine protein sources	14.5
Marine oils	10.4
Other	14.5
Total	100

Table 2.
Norwegian salmon feed ingredients used in 2016 (values in percentage %). In line with the need to reduce exploitation of marine fish meal, ingredients for feed formulation now comprise of 40.2% plant protein sources, while marine protein sources are reduced to about 14.5%.

protein content, ease of availability and affordability [45]. Other sources of plant ingredients for possible use in formulating diets for salmon include wheat glutten, barley, pea, lupin, corn maize, sunflower, linseed, olive and palm oil. Similarly, vegetable oil is a suitable replacement for fish oils in formulation of feeds for salmon [48]. However, proper attention is required in the choice of the plant material, to ensure that it meets the required amounts of protein, the high amounts of starch in plant matter is adequately reduced, meets suitable profiles for amino acids and minerals, as well as reduced levels of fiber and anti-nutritional factors [45].

The ingredients that constituted the largest portion in Norwegian salmon feed was soy protein content which was 19% and rapeseed oil together with camelina oil accounted for 19.8% while wheat and wheat glutten accounted for 17.9% [44]. The ingredients used in Norwegian salmon feeds in 2016 are as shown in the table below (**Table 2**).

5. Genetics and genomics to support improved breeding of salmon

Salmon is iconic not only in its ecology, life cycle, ability to oscillate among different environments, high conservation value, but also in its genomic organization. As tetraploid individuals, the genome of salmon evolved through a historical autotetraploidization whole genome duplication (WGD) [49], which occurred 88–103 million years ago [50]. Autotetraploidization occurs by a spontaneous doubling of all chromosomes [5], creating four pairs of chromosomes that recombine spontaneously during meiosis after WGD. Like the normal diploid gametes, there is a reduction in the ploidy state of salmon genome (halfing), a rediploidization process that returns the salmon genome to diploid state prior to recombination [5]. Enormous structural re-organization occurs in the salmon genome during rediploidization, with some parts of the genome remaining tetraploid [51], mismatch in recombination rates of females and males [51], and the retention of half of the genes of the species in duplicated state from the salmonid specific 4th round (Ss4R) of WGD [52]. Apart from this reorganization of the salmonid genome during rediploidization, which creates suitable substrates for the evolution of salmonids, a fifth of salmon genes retained a pair of more ancient gene duplicates from the Teleost specific 3rd round of WGD (Ts3R) [5]. This increases the diversity and complexity of gene families in salmonids, compared to other teleost fishes, which increases evolutionary potential as well as heritability and genetic potential during selective breeding of salmon for commercial aquaculture. The overall effect of these events is a much higher variability in the gene pool, from which samples for generating F1 are drawn.

5.1 Selective breeding in salmon aquaculture

Selective breeding in support of salmon aquaculture began in Norway, a lead producer of *S. salar* in the 1990s, with the first such programme initiated in 1997, using a total of 40 strains collected across rivers country wide [53]. Concerted efforts produced 4 more strains: Mowi, Rauma, Jakta and Bolaks strains [54], which have been crossed and used extensively, including export to other salmon growing countries. The breeding programme focused on growth rate, with a substantially superior genetic gain per generation of 15% being achieved. This rate is comparatively better than tilapias, where for instance the GIFT strain achieved genetic gain of 12–17% in the fifth generation, compared to 15% in the first generation of salmon [55]. Similarly, in China, the ProGift strain of *Oreochromis niloticus* reported a genetic gain of 11.4% in the 6th generation [56], translating to increased growth of 60–90% bigger body weight at harvest [56]. The high rate of genetic gain in salmon could be attributed to selection intensity, recent history of domestication, in addition to a complex genome following whole genome duplication events [5].

With improved technology and changing interests of salmon breeders in the 1990s, the breeding objectives, moved from growth rate to other complex challenges, such as disease resistance, rationalized by increased incidences of infectious pancreatic necrosis virus (IPNV) for instance [5]. Indeed, disease outbreaks are a major challenge in farmed salmon production in some lead producer countries. To this end, marker assisted selection helped identify individuals with QTL for higher resistance to IPNV [57], resulting to reduced incidence of IPNV, and therefore better yields.

5.2 Genetic mapping

One of the major challenges facing intensive farming of salmonids is infectious diseases, which often occasion huge losses to farmers, and slows down the rate of expansion of salmon farming. Most of these diseases are caused by bacteria, viruses and parasites [58], whose severity and frequency of occurrence increases with the level of intensification of production. Infectious salmon anemia (ISA), Infectious pancreatic necrosis (IPN), Skeletal muscle inflammation (HMSI), and pancreas disease are viral diseases of salmon [36, 58], which also lower growth rates and increase costs of treatment [58]. The main bacterial disease is the salmon rickettsial syndrome, which causes huge economic losses, while the sea lice disease is the main parasitic disease in farmed salmon. On the other hand, the amoebic gill disease is the main protozoan disease in farmed salmon [59], which also increases susceptibility to other infections. Most of the conventional preventive and prophylactic measures used to control these diseases such as vaccination, antibiotics and anti-parasitic drugs, or biosecurity [58], are often not effective. To counter these losses and increase economic returns of salmon farming ventures, selective breeding for resistance to diseases is often applied, based mainly on information from relatives (sib information) [58], since the trait is difficult to measure directly on candidate fish for selection.

5.3 Breeding for disease resistance in farmed salmon

Global growth of aquaculture is often constrained by progressive loss of quality of the breeding germplasm due to inappropriate fish husbandry as well as selection requisite in particular fish breeding schemes and the repetitive use of certain (good looking or higher yielding) brood stock, and incidences of diseases, especially as production is intensified in pursuit of food security, higher incomes and livelihood.

Disease resistant fish are those that limit infection by curtailing the replication of the pathogen in the body of the fish [60]. In itself, disease resistance is a precious trait in fish, animal or plant breeding programmes, for it limits wanton use of chemicals or drugs, whose effect is more-broad based, even to non-target organisms in the environment [61], yet their efficacy at limiting incidences and severity of diseases may not be sufficient. Similarly, resistance to drugs or antibiotics by microbes is a real and serious problem in agricultural production [62], exacerbated by global warming [62]. Therefore, alternative, more environment friendly, cost effective and sustainable approaches are desirable in controlling diseases in salmon aquaculture. One of these strategies is the breeding of superior strains, which exploits natural genetic variation for disease resistance to improve the quality, efficiency, profitability and sustainability of the aquaculture enterprise. Today, breeding for improved strains or varieties is a highly efficient process, because of the increasing tool kit of genomic resources, especially for the high throughput next generation sequencing technologies. Therefore, it has been possible to focus on growth, sex determination or disease resistance as breeding objectives [63]. In farmed salmon production, breeding for disease resistant strains is an active agenda since the 1990s [53], since it imparts cumulative and permanent resistance to diseases in the fish. Breeding for resistance nevertheless requires a population of sufficient genetic variation for the trait. High levels of additive genetic variation for disease resistance are reported in different salmonid species (**Table 3**), indicating possibility of deriving gains in selective breeding for disease resistance in salmonids.

Therefore, it is possible to improve resistance to diseases in salmonids through genetic improvement, as a tool in disease control in salmon aquaculture,

Species	Pathogen	Heritability (h^2 ± S.E)	Reference
Salmo salar	*Renibacterium salmoninarum*	0.2 ± 0.1	[64]
	Aeromonas salmonicida	0.48 ± 0.17	[65]
	A. salmonicida	0.59 ± 0.06	[66]
	A. salmonicida	0.62	[67]
	IPNV	0.55	[67]
	Infectious Salmon Anemia Virus (ISAV)	0.13 ± 0.03	[68]
	ISAV	0.16 ± 0.01	[69]
	ISAV	0.24 ± 0.03	[66]
	ISAV	0.37	[67]
	Vibrio anguillarum	0.38 ± 1.07	[68]
	Vibrio salmonicida	0.13 ± 0.08	[64]
	Caligus royercresseyi	0.10 ± 0.03	[58]
	Piscirickettsia salmonis	0.18 ± 0.03	[58]
Salvelinus fontinalis	*A. salomnicida*	0.51 ± 0.03	[70]
O. mykiss	*Yersinia ruckeri*	0.21 ± 0.05	[71]
	Flavobacterium psychrophilum	0.35 ± 0.09	[72]
	F. psychrophilum	0.07 ± 0.02	[71]
	Viral heamorrhagic septicemia	0.11 ± 0.1	[71]

Table 3.
Heritability for resistance to different infectious and parasitic diseases in salmonid species. Adopted from [58].

since heritability for disease resistance is high (**Table 3**) [64–72]. Typically, disease resistance in salmon has been determined through marker assisted selection or genomic selection based only on information from relatives, since it is very difficult to measure disease resistance in the actual fish. By this approach, it is difficult to determine the genetic gain per generation imparted to the fish individuals by the selection effort [58]. This slowed the rate at which selection of disease resistant fish individuals and the realization of highly resistant individuals progresses, since estimated breeding values from sib information is less accurate than would that from the selection candidates themselves [58]. Previous research efforts determined correlation between immune parameters and resistance to diseases in salmon [73]. However, while this may be a pointer to some of the fish individuals that may be resistant to diseases, the total variability in survival of salmon is too low to be attributed to immune variables [58]. Similarly, resistance of fish to diseases is a function of many more factors, and not just immune parameters. Although high genetic variability necessary for improvement of disease resistance exists in salmon, correlations between genetic variation and disease resistance report mixed results [58], with non-existent relationship [72], negative relationship [74], or low to moderately positive relationship [70].

Due to these complexities in studying disease resistance for breeding improved strains for commercial production of salmon, genomic resources have been developed over the last decade, to enable a more focused approach to breeding for disease resistance in salmon. These include: high quality reference sequence for trout, which is also applicable to salmon [52], high density SNP genotyping arrays for *S. salar* [75], and lower density SNP platform for QTL mapping [76]. These resources support the study and understanding of the genetic basis of disease resistance in salmon through identification of candidate genes for resistance to certain diseases, mapping QTL regions with genes of interest for resistance to certain diseases, and gene expression studies [58] in fish challenged with certain pathogens.

5.4 Studying candidate genes driving disease resistance in salmonids

This approach of understanding disease resistance in aquaculture species exploits the candidate gene theory, in which phenotypic variance for a trait in a population is a result of polymorphisms that exist in genes known to drive that trait [77], and utilizes annotated gene sequences of known function [58]. Due to limited availability of annotated gene sequences in most aquaculture species, studies of association between candidate genes and resistance to diseases has shifted to the Major Histocompatibility Complex (MHC) [58]. The MHC is a multigene family, or a gene-complex region, comprising several genes mediating diverse immune and phenotypic responses or characteristics [78], and interfaces the immune system and pathogens [78]. The MHC presents the class I and II genes, which encode polypeptides that recognize and bind self and foreign peptides and present them to T-cells for destruction [79]. The MHC class I genes bind peptides produced by intracellular degradation of pathogens (such as viruses), and present them to the immune system (cytotoxic T-cells), triggering cellular immune response that destroys the cells. On the other hand, class II genes bind peptides produced outside cells (e.g. bacteria) and present them to helper T-cells, which secrete cytokine mediators. Cytokines elicit humoral (antibody), cytotoxic and inflammatory responses that destroy the pathogens. A unique feature of MHC gene complex is its high levels of polymorphism, with different regions showing high allelic diversity [78]. For instance, *S. salar* from the Baltic Sea has a single MHC class IIB locus with up to 16 alleles within populations [80]. This diversity, thought to be maintained by balancing selection in different taxa, is what makes the MHC a hotbed of scientific interest and research. Class I and class II genes are well characterized and highly polymorphic in *S. salar*, and rainbow trout. Association between MHC class IIB alleles and resistance against *A. salmonicida* is reported [81], while variant fish for MHC class I and II are susceptible to IHN [79], but have resistance to furunculosis and ISA [82]. Some salmon fish individuals that bear certain genes in the MHC are more susceptible to furunculosis [81]. These studies seem to suggest that a clear understanding of the MHC and its associated polymorphism can provide useful insights in selecting suitable phenotypes of salmon for breeding for disease resistance, to support intensive and commercial production of the fish. While the number of studies showing correlation of MHC genes to disease resistance and vice versa in salmonids is on the rise, these largely form anecdotal evidence rather than solid evidence for correlation between certain genes of the MHC and resistance to diseases in salmon. This is because resistance to diseases in salmon is a polygenic trait, driven by several genes rather than certain gene(s). Furthermore, the class I genes in the MHC are highly diverse, and this large number of alleles seems to mask the effect of certain alleles, making it difficult to study roles of such alleles in disease resistance or susceptibility. However, since disease resistance traits are typically polygenic, future efforts to understand the genetic basis of disease resistance in salmon should study genetic architecture of variants in the whole genome, as well as possible interactions between genes. Additionally, for populations of salmon where correlation between certain genes and disease resistance or susceptibility has been demonstrated, even where such correlation only seems anecdotal, research should concentrate on studying the suitability of such populations (genotypes) as brood stock for seeds used in selection to improve resistance to diseases. Additionally, loci already identified as having some association with disease resistance should be tested further using modern marker technology, such as next generation sequencing, to improve the confidence of inference.

5.5 Mapping QTL regions for resistance to diseases in salmon

Quantitative trait locus (loci) (QTL), is the variability of loci, leading to increased variation in the expression of a quantitative character [83]. A QTL is a locus that controls a quantitative phenotypic trait, identified by showing a statistical association between genetic markers surrounding the locus and phenotypic measurements [84]. The presence of QTL improves the understanding of the number of genes and their relative effects in determining expression of the trait. The identified QTL is then mapped through marker association (association mapping) in the whole genome, thereby identifying genomic regions involved in genetic variation of a trait. Fish individuals with the identified QTL or the genomic regions are used in the breeding programme if the QTL is of advantage, or left out if the QTL confers a disadvantage to the fish. SNP markers are especially important in the construction of high-density maps, which are used to fine map QTLs and facilitate identification of causative genes involved in genetic variation for specific characters. SNP markers are available for salmonids [85], and are used in high resolution mapping of disease resistance genes. Since QTL mapping relies on molecular markers, the technique is likely to be used in many breeding schemes, due to the presence of many modern marker technologies, most of which increase the throughput and subsequent output. In this regard, marker technologies like Genotyping by Sequencing (GBS) are already being used in salmon breeding [86]. Coupled with technologies like the Genome wide association studies, these next generation sequencing platforms are likely to accelerate breeding disease resistant salmon strains for use by farmers in commercial aquaculture. These highly versatile NGS platforms enabled the construction of genetic linkage maps, some incorporating several different markers. GBS has especially opened up new opportunities for genotyping SNPs, which are used to construct dense linkage maps [87], from which genes driving commercially important traits like disease resistance can be deciphered, to aid choice of desirable genotypes for use in the breeding schemes by farmers. For instance, meiotic maps have been developed for sockeye salmon, *O. nerka* [87], suitable for salmonids as tetraploid fishes having duplicated genomes, and which enable comparative genomics and association mapping of important genes or genomic regions of importance in the fish breeding schemes. Analysis of genetic linkage maps aids the location of QTL or genes for important traits for aquaculture production.

In comparative genomics, genomic features in complete genome sequences of different organisms or species are compared. These genomic features vary from DNA sequences, genes, gene orders, regulatory sequences to other genomic structural land marks, that distinguish fish individuals, and can therefore be used to identify suitable genotypes (by comparing regions of similarity and differences) for use in breeding schemes for profitable aquaculture. Autotetraploidization of the common salmonid ancestor 50–120 million years ago [49] made salmonids iconic in character, value and evolutionary potential. As a tetraploid resulting from a duplicated genome therefore, sex determination in salmonids is one of the most complex traits, and probably represents a classic example of diversification in this group of fishes. In itself, tetradiploidization provided evolutionary pressure to diversify species in the Salmonid family in to 11 genera [1]. In this regard, it is interesting to know if sex determination, a trait important for aquaculture, is influenced by the same mechanism for each of the genera within the family, or indeed different mechanisms underpin the process in different genera, or whether different species have different mechanisms, or whether the sex determining gene is the same in the different species, or has shifted to a different chromosome in different species, or whether sex has evolved differently in the different salmonid species that radiated following tetra ploidy. These are uncertainties that have been addressed by comparative mapping, where linkage maps developed for different species are compared, to study the presence or absence of genetic elements of interest, or if these genetic elements are located within different linkage groups or on different chromosomes. Through these comparisons, suitable genotypes are isolated for breeding for the trait of interest.

Through concerted efforts to generate linkage maps for species of salmon, many studies now report sex determining locus on the end of the long arm of chromosome 2. In other species like the brown trout, karyotypic studies report absence of sex determining chromosomes [88]. With this kind of information, gleaned from genetic linkage maps, suitable genotypes can be selected, which when crossed have very high chance of producing monosex seeds, either male or female, depending on which sex is preferred by the farmer. Due to enormous power of linkage mapping in identifying QTLs with important roles in tolerance or resistance of salmonids to diseases that limit intensive and profitable culture of salmon, several such maps have been developed for rainbow trout [89], *S. salar*, *Salmo trutta* [90], the Arctic trout, *Salvelinus alpinus* [91], and Sockeye salmon, *S. nerka* [87]. Apart from these linkage maps from which QTL for tolerance or resistance of salmon for diseases have been inferred, several other studies have also been carried out to improve breeding for disease resistance. For instance, QTL in a back cross of strains of rainbow trout resistant and susceptible to IPN has been detected [92], while QTL for resistance to whirling disease in rainbow trout has been detected [93]. Similarly, with these maps, much more information has been gleaned, such as on which chromosome sex determining genes are located, conservation of synteny, rates of recombination in certain species, loss of genes following autotetraploidization events among salmonids and the rate at which such genes were lost, or indeed the emergence of new genetic or sequence features among different salmonid species that radiated following tetraploidization. These research efforts, among a majority of other and ongoing studies, are being incorporated in to breeding schemes by salmon farmers, and demonstrate the importance of salmonids both as high value fish food and sentinel species [94] for human consumption and conservation respectively, and partly explain why farmed salmon production is always on the increase.

On the other hand, association mapping or linkage disequilibrium mapping is the linking of observed phenotypes to the presence of the genotype which drives the observed phenotypic characters or variations [95]. The farmer is interested in seeing the best phenotype from the fish stocked in the ponds or cages or ranches that form his farm, as this translates to higher average tonnage of fish produced and therefore sufficient food for consumption, export and higher profitability of the enterprise. Therefore, to help farmers sustain profitability of their enterprises, researchers try to apply linkage disequilibrium mapping to choose the best genotypes of salmonid species which when used by farmers, give the highest produce, with respect to the trait that the farmer is interested in. As the very basis of Marker assisted breeding, association mapping has hastened identification of suitable genotypes for use in breeding schemes for salmon, addressing a specific breeding objective. In this regard, using RFLP markers, association is reported in backcross families resistant and susceptible to IHN [96]. This means backcross genotypes with RFLP markers show resistance to IHN, while those without RFLP markers are susceptible to IHN, and this easily guides which fish individuals to choose for use in the breeding scheme for resistance against IHN, and which rainbow trout fish individuals to discard, because they lack resistance to IHN. Similarly, AFLP markers associate with resistance to ISA in two full-sib families of *S. salar* [97], with the AFLP markers mapped on to linkage group 8 of the *S. salar* genome [98].

The efforts at comparative mapping and linkage disequilibrium mapping for salmon will be more relevant for precision or more efficient breeding of better performing salmon strains when a large number of markers are available on the map in a clear order, supported by testing of a large number of families (both half sib and full sib) for the presence or lack of important QTLs that associate with certain phenotypes. The studies enumerated above appear disjointed, but will fit better in the international collaboration to sequence the Atlantic salmon genome to be used as the reference sequence for many salmonid species [86] for improved breeding. Furthermore, as more efficient and cost-effective next generation sequencing platforms become available, both genetic high density SNPs marker panels and high-resolution genetic maps are generated, from which association between markers and QTLs for important traits are deciphered [94]. Mapping of these genetic resources on to the reference map generated for *S. salar* [94] facilitates the identification of mutations [58] that underpin disease resistance in many salmonid species and guide precision breeding for improved resistance to diseases for more profitable aquaculture enterprises.

5.6 A study of gene expression for disease resistance

One of the ways fish resist diseases and therefore some infections do not translate in to full scale sickness is through mounting an immune response to the infection or the pathogen. Related to this immune response are a series of mechanisms that synergize the protective apparatus of the fish. In a population of fish, there will be some individuals with a higher ability to resist pathogens and therefore full manifestation of sickness or symptoms (i.e. resistant fish), and fish that lack ability or do not have sufficient ability to resist pathogens, and therefore develop the disease (i.e. susceptible fish). Therefore, functional genetic variation for disease resistance will exist in such population of fish [58]. Fish that are resistant to diseases usually show less infection, with fewer viral or bacterial pathogens getting in to the body or cells of the fish. This is because differential immune response in resistant fish inhibits attachment of the pathogen, as well as entry and subsequent replication of the pathogen in fish cells. This inhibition allows the immune system of the fish ample time to mount a sufficient response to completely combat the pathogens [91]. However, should the pathogen succeed to enter the body of the fish, several genes are released by the fish to help fight off foreign invasive agents, with a faster rate of and more intense release in susceptible than resistant fish [99].

For IPNV infection in salmon for instance, interferon induced genes are released to fight the virus. These include Mx, ISG 15, Vip-2, gig 2, and CCL 19 [100]. Additional genes activated to fight IPNV infection in salmon include: Interferon regulatory factor 3 and 8, interleukin 3 receptor, and the macrophage colony stimulating factor, transcription factor 3, and the transcription factor E2-alpha [99]. Similarly, infection with furunculosis stimulates gene expression in response, to help counter-attack the infection. In this regard, Mx1, ubiquitin-protein ligase HERC 4, HERC 5 and HERC 6, ISG 15, eukaryotic translation Initiation factor 4 gamma 1 are elicited during an invasion with ISAV [101]. On the other hand, furunculosis infection in salmon elicits several genes, including JunC, JunD, NFkB, NF-kappaB-105, NFkB1, CYP3A4 and fibronectin [102], which act in different ways and mechanisms to confer protection of the fish against the effects of furunculosis. Since these genes are expressed differently in resistant and susceptible fish, and the fact that in a population of salmon, there are disease resistant and susceptible individuals, resistance has a genetic basis [99, 101, 102]. Therefore, gene expression profiles for disease resistance and susceptibility can help identify suitable fish for use in the breeding programme to improve the resistance of the fish to diseases.

In conclusion, future efforts should focus more on saturating the genetic map for salmon, from which loci for commercially important traits can be inferred, to support breeding efforts for improved strains.

Conflict of interest

Authors declare no conflict of interest.

Author details

James E. Barasa*, Purity Nasimiyu Mukhongo and Cynthia Chepkemoi Ngetich
Department of Fisheries and Aquatic Sciences, University of Eldoret, Kenya

*Address all correspondence to: barkanoti@gmail.com

References

[1] Nelson JS. Fishes of the World. 3rd ed. New York: Wiley and Son; 2006

[2] Criddle KR, Shimizu I. The economic importance of wild Pacific salmon. In: PTK W, Noakes DJ, editors. Salmon. Hauppauge, New York, USA: Nova Science Publishers Inc.; 2014. Chapter 14. 39 pages. ISBN: 978-1-63117-570-1

[3] Weitkamp LA, Goulette G, Hawkes J, O'Malley M, Christine L. Juvenile salmon in estuaries: Comparisons between North American Atlantic and Pacific salmon populations. Reviews in Fish Biology and Fisheries. 2014;**24**(1). DOI: 10.1007/s11160-014-9345-y

[4] Food and Agriculture Organization of the United Nations (FAO). The State of World Fisheries and Aquaculture 2020. Sustainability in action. Rome: FAO; 2020. DOI: 10.4060/ca9229en.

[5] Houston RD, Macqueen DJ. Atlantic salmon (*Salmo salar* L.) genetics in the 21st century: Taking leaps forward in aquaculture and biological understanding. Animal Genetics. 2018;**50**:3-14. DOI: 10.1111/age.12748

[6] Bergheim A, Fivelstad S. Atlantic salmon, *Salmo salar* in aquaculture: Metabolic rate and water flow requirements. In: Nook PTJ, Noakes DJ, editors. Salmon: Biology, Ecological Impact and Economic Importance. 1st ed. Hauppauge, New York, USA: Nova Publishers Inc.; 2014. Chapter 8. 20 pages

[7] Dresdner J, Chavez C, Estay M, Gonzalez G, Salazar C, Santis O, et al. Evaluacion socioecon omica del sector salmonicultor, en base a las nuevas exigencias de la Ley General de Pesca y Acuicultura. Informe Final. Proyecto FIPA 2015-42. Chile: Fondo de Investigacion Pesquera y de Acuicultura, Ministerio de Economıa; 2016. p. 351

[8] Avendano HR. Proper antibiotics use in the Chilean salmon industry: Policy and technology bottlenecks. Aquaculture Reports. 2018;**495**:803-805

[9] Buschmann AH, Cabello F, Young K, Carvajal J, Varela DA, Henrıquez L. Salmon aquaculture and coastal ecosystem health in Chile: Analysis of regulations, environmental impacts and bioremediation systems. Ocean and Coastal Management. 2009;**52**:243-249. DOI: 10.1016/J.ocecoaman.2009.03.002

[10] Lepe-Lopez M, Escobar-Dodero J, Zimin-Veselkoff N, Azat C, Mardones FO. Assessing the present and future habitat suitability of *Caligus rogercresseyi* (Boxshall and Bravo, 2000) for salmon farming in Southern Chile. Frontiers in Veterinary Science. 2021;7:61539

[11] Hevroy EH, Hunoskar C, de Gelder S, Shiminzu M, Waaggloo R, Breck O, et al. GH-IGH system regulation of attenuated muscle growth and lipolysis in Atlantic salmon, reared at elevated sea temperature. Journal of Comparative Physiology. 2013;**183**(2):243-259

[12] INGELAND. Levantamiento de informacion de piscicul- turas en chile y su incorporacion a la IDE de la Divisi on de Acuicultura. Informe Final Proyecto FIPA N° 2016-19. Chile: Fondo de Investigacion Pesquera y de Acuicultura, Ministerio de Economıa; 2017. p. 109

[13] Quinones RA, Fuentes M, Montes RM, Soto D, Leon-Munoz J. Environmental issues in Chilean salmon farming: A review. Reviews in Aquaculture. 2019;**11**:375-402. DOI: 10.1111/raq.12337

[14] Buschmann AH, JLP M. Challenges for future salmonid farming. In: Cochran JK, Bokuniewicz JH, Yager LP, editors. Encyclopeadia of Ocean Sciences. 3rd ed. Amsterdam, The Netherlands: Elsevier; 2019. pp. 313-319. DOI: 10.1016/B978-0-12-409548-9.09560-4

[15] Rypel AL, David RS. Pattern and scale in latitude–production relationships for freshwater fishes. Ecosphere. 2017;**8**(1):e01660

[16] Dolbeth M, Cusson M, Sousa R, Pardal M, Prairie YT. Secondary production as a tool for better understanding of aquatic ecosystems. Canadian Journal of Fisheries and Aquatic Sciences. 2012;**69**:1230-1253

[17] Liltved H, Homme M, Gangnes SSL, Hess-Erga OK, Uleberg S, Drengstig A. Can aquaponic systems be adapted for Norwegian aquaculture. Fiskehelse. 2012:14-18. ISBN 978-82-8208-028-6

[18] Badiola M, Mendiola D, Bostock J. Recirculating aquaculture systems (RAS) analysis: Main issues on management and future challenges. Aquaculture Engineering. 2012;**51**:26-35. DOI: 10.1016/j.aquaeng.2012.07.004

[19] Blancheton JP, Piedrahita R, Eding EH, Roque d'Orbcastel DE, Gilles L, Bergheim A, et al. Intensification of land-based aquaculture production in single pass and reuse systems. Aquacultural Engineering and Environment. 2007:21-47

[20] Badiola M, Basurko O, Gabiña G, Mendiola D. Integration of energy audits in the life cycle assessment methodology to improve the environmental performance assessment of recirculating aquaculture systems. Journal of Cleaner Production. 2017;**157**:155-166. DOI: 10.1016/j.jclepro.2017.04.139

[21] Clark R, Bostock J. Regional review on Status and Trends in Aquaculture in Europe-2015. Rome, Italy: Food and Agriculture Organization of the United Nations (FAO); 2007. FAO Fisheries and Aquaculture Circular No. 1135/1 FIAA/ C1135/1 (En). 52 pages

[22] Liu Y, Rosten TW, Henriksen K, Hognes ES, Summerfelt S, Vinci B. Comparative economic performance and carbon footprint of two farming models for producing Atlantic salmon (*Salmo salar*): Land-based closed containment in freshwater and open net in sea water. Aquacultural Engineering. 2016;**71**:1-12. DOI: 10.1016/j.aquaeng.2016.01.001

[23] Colt J, Summerfelt S, Pfeiffer T, Fivelstad S, Rust M. Energy and resource consumption of land-based Atlantic salmon smolt hatcheries in the Pacific Northwest (USA). Aquaculture. 2008;**280**:94-108

[24] Morrisey M. Development of value-added products in aquaculture. In: Cruz-Suárez LE, Ricque-Marie D, Tapia-Salazar M, Nieto-López MG, Villarreal-Cavazos DA, Gamboa-Delgado J, Hernández-Hernández L, editors. Avances en Nutricion Acuicola; 23-25 de Noviembre; san Nicolas de los Garza, N. L. Mexico. Universidad Autónoma de Nuevo, León, Monterrey, México; 2011. pp. 12-27. ISBN 978-607-433-775-4

[25] Hofmann T. Recirculating aquaculture system-based salmon farming. Field Actions Science Reports. 2019;**20**. Available from: https://journals.openedition.org/factreports/5766

[26] Binsi PK, Sreelakshmi KR, Ashok KK. Value added fish products. In: Suresh A, Sajeev MV, Rejula K, Mohanty AK, editors. Extention Management Techniques for Up-scaling Technology Dissemination in Fisheries (e manual). Cochin, India: Central Institute of Fisheries Technology; 2018. pp. 47-57

[27] Alam N. Packaging of fish and fish products. In: Alam NMK, editor. Participatory Training of Trainers: A New Approach Applied in Fish Processing. Dhaka, Bangladesh: Bangladesh Fisheries Research Forum. 2007. p. 329

[28] Hastein T, Lindstad T. Diseases in wild and cultured salmon: Possible interaction. Aquaculture. 1991;**98**:277-288

[29] Naylor RL, Hardy RW, Buschmann AH, Bush SR, Cao L, Klinger DH, et al. A 20-year retrospective review of global aquaculture. Nature. 2021;**591**:551-574

[30] Godoy MG, Kibenge MJT, Suarez R, Lazo E, Heisinger A, Aguinaga J, et al. Infectious salmon anaemia virus (ISAV) in Chilean Atlantic salmon (Salmo salar) aquaculture: Emergence of low pathogenic ISAV HPR0 and re-emergence of virulent ISAV- HPRΔ: HPR3 and HPR14. Virology Journal. 2013;**10**(1):344-361

[31] Iversen A, Asche F, Hermansen O, Nystoyl R. Production cost and competitiveness in major salmon farming countries 2003-2018. Aquaculture. 2020;**522**:735089

[32] Lyngstad TM, Qviller L, Sindre H, Brun E, Kristoffersen AB. Risk factors associated with outbreaks of infectious salmon anemia with unknown source of infection in Norway. Frontiers in Veterinary Science. 2018;**5**:308. DOI: 10.3389/fvets.2018.00308

[33] Vågsholm I, Djupvik HO, Willumsen FV, Tveit AM, Tangen K. Infectious salmon anaemia (ISA) epidemiology in Norway. Preventive Veterinary Medicine. 1994;**19**:277-290. DOI: 10.1016/0167-5877(94)90095-7

[34] Mardones FO, Martinez-Lopez B, Valdes-Donoso P, Carpenter TE, Perez AM. The role of fish movements and the spread of infectious salmon anemia virus (ISAV) in Chile, 2007-2009. Preventive Veterinary Medicine. 2014;**114**:37-46. DOI: 10.1016/j.prevetmed.2014.01.012

[35] Gustafson L, Remmenga M, Sandoval del Valle O, Ibarra R, Antognoli M, Gallardo A, et al. Area contact networks and the spatio-temporal spread of infectious salmon anemia virus (ISAV) in Chile. Preventive Veterinary Medicine. 2016;**125**:135-146. DOI: 10.1016/j.prevetmed.2016.01.001

[36] Roberts RJ, Pearson MD. Infectious pancreatic necrosis in Atlantic salmon, *Salmo salar* L. Journal of Fish diseases. 2005;**28**:383-390

[37] Smail DA, Bruno DW, Dear G, Mcfarlane LA, Ross K. Infectious pancreatic necrosis (IPN) virus species serotype in farmed Atlantic salmon, *Salmo salar* L. post-smolts associated with mortality and clinical disease. Journal of Fish Diseases. 1992;**15**:77-83

[38] Melby HP, Krogsrud J, Hastein T, Stenwig H. All commercial Atlantic salmon seawater farms in Norway harbour carriers of infectious pancreatic necrosis virus (IPNV). In: Fryer JL, editor. 2nd International Symposium on Viruses of Lower Vertebrates. Corvallis, O.R: Oregon State University; 1991. pp. 211-217

[39] Christie KE. Immunization with viral antigens, infectious pancreatic necrosis. In: Gudding R, Lillehaug A, Midtlying PJ, Brown F, editors. Fish Vaccinology. 1997. pp. 191-199

[40] Skjesol A, Aamo T, Hegseth MN, Robertsen B, Jørgensen JB. The interplay between infectious pancreatic necrosis virus (IPNV) and the IFN system: IFN signaling is inhibited by IPNV infection. Virus Research. 2009;**143**:53-60

[41] Bruno DW. Furunculosis. Revised edition. ICES Identification leaflets for diseases and parasites of fish and shellfish. Leaflet No. 37:1-5. ISBN 978-87-7482-172-4. DOI: 10.17895/ices.pub.52

[42] Jarp J, Tangen K, Willumsen MV, Djupvik HO, Tveit AM. Risk factors for infection with *Aeromonas salmonicida* subspecies *salmonicida* in Norwegian Freshwater hatcheries. Diseases of Aquatic Organisms. 1993;**17**:81-86

[43] Oliveira VHS, Dean KR, Qviller L, Kirkeby C, Jensen BB. Factors associated with baseline mortality in Norwegian

Atlantic salmon farming. Scientific Reports. 2021;**11**:14702

[44] Aas TS, Ytrestoyl T, Asgard T. Utilization of feed resources in the production of Atlantic salmon *Salmo salar*, in Norway: An update for 2016. Aquaculture Reports. 2019;**15**:100216. DOI: 10.1016/j.aqrep.2019.100216

[45] Gillund F, Myhr AI. Perspectives on salmon feed: A deliberative assessment of several alternative feed resources. Journal of Agricultural and Environmental Ethic. 2010;**23**:527-550

[46] Naylor RL, Hardy RW, Bureau DP, Chiu A, Elliott M, Farrell AP, et al. Feeding aquaculture in an era of finite resources. Proceedings of the National Academy of sciences. 2009;**106**:15103-15110

[47] Nicol SN, Foster J. Recent trends in the fishery for Antarctic krill. Aquatic Living Resources. 2003;**16**:42-45

[48] Turchini GM, Torstensen BE, Wing-Keong NG. Fish oil replacement in finfish nutrition. Reviews in Aquaculture. 2009;**1**:10-57

[49] Allendorff FW, Thorgaard GH. Tetraploidy and the evolution of salmonid fishes. In: Turner BJ, editor. Evolutionary Genetics of Fishes. New York, NY: Plenum Press; 1984. pp. 1-53

[50] Macqueen DJ, Johnston IA. A well-constrained estimate for the timing of the salmonid whole genome duplication reveals major decoupling from species diversification. Proceedings of the Royal Society B: Biological Sciences. 2014;**281**:20132881

[51] Allendorf FW, Bassham S, Cresko WA, Limborg MT, Seeb LW, Seeb JE. Effects of crossovers between homeologs on inheritance and population genomics in polyploid-derived salmonid fishes. Journal of Heredity. 2015;**106**:217-227

[52] Berthelot C, Brunet F, Chalopin D. The rainbow trout genome provides novel insights into evolution after whole-genome duplication in vertebrates. Nature Communications. 2014;**5**:3657

[53] Gjøen H, Bentsen H. Past, present, and future of genetic improvement in salmon aquaculture. ICES Journal of Marine Sciences. 1997;**54**:1009-1014. DOI: 10.1016/S1054-3139(97)80005-7

[54] Glover KA, Solberg MF, McGinnity P, Hindar K, Verspoor E, Coulson MW, et al. Half a century of genetic interaction between farmed and wild Atlantic salmon: Status of knowledge and unanswered questions. Fish and Fisheries. 2017;**18**:890-927

[55] Eknath AE, Dey MM, Rye M, Gjerde B, Abella TA, Sevilleja R, et al. Selective breeding of Nile tilapia for Asia. In: Proceedings of the 6th World Congress on Genetic Application in Livestock Production. Vol. 27. Amidale, Australia: University of New England; 1998. pp. 89-96

[56] Thodesen J, Rye M, Wang YX, Yang KS, Bentsen HB, Gjedrem T. Genetic improvement of tilapias in China: Genetic parameters and selection response in growth of Nile tilapia after 6 generations of multi-trait selection for growth and fillet yield. Aquaculture. 2011;**322-323**:51-64

[57] Houston RD, Haley CS, Hamilton A, Guy DR, Tinch AE, Taggart JB, et al. Major quantitative trait loci affect resistance to infectious pancreatic necrosis in Atlantic salmon (*Salmo salar*). Genetics. 2008;**178**:1109-1115

[58] Yanes JM, Houston DR, Newmann S. Genetics and genomics of disease resistance in salmonid species. Frontiers in Genetics. 2014;**5**(415):1-13

[59] Mitchell SO, Rodger HD. A review of infectious gill disease in marine

salmonid fish. Journal of Fish Diseases. 2011;**34**(6):411-432. DOI: 10.1111/j.1365-2761.2011.01251.x

[60] Doeschl-Wilson AB, Villanueva B, Kayriazakis I. The first step towards genetic selection for host tolerance to infectious pathogens: Obtaining the tolerance phenotypes through group estimates. Frontiers in Genetics. 2012;**3**:265. DOI: 10.3389/fgene.2012.00265

[61] Caipang CMA, Suharman I, Avillanosa AL, Bargoyo VT. Host derived probiotics for finfish aquaculture. Earth and Environmental Science. 2020;**430**:1-11

[62] Rodriguez-Verdugo A, Gaut BS, Tenaillon O. Evolution of *Escherichia coli* rifampicin resistance in an antibiotic-free environment during thermal stress. Molecular Biology and Evolution. 2013;**33**:25-39

[63] Robledo D, Palaiokostas C, Bargellon L, Martinez P, Houston R. Applications of genotyping by sequencing in aquaculture breeding and genetics. Reviews in Aquaculture. 2018;**10**:670-682. DOI: 10.1111/raq.12193

[64] Gjedrem T, Gjøen H. Genetic variation in susceptibility of Atlantic salmon, *Salmo salar* L., to furunculosis, BKD and cold water vibriosis. Aquaculture Research. 1995;**26**:129-134. DOI: 10.1111/j.1365-2109.1995.tb00892.x

[65] Gjedrem T, Salte R, Gjøen HM. Genetic variation in susceptibility of Atlantic salmon to furunculosis. Aquaculture. 1991;**97**:1-6. DOI: 10.1016/0044-8486(91)90274-B

[66] Olesen I, Hung D, Ødegård J. Genetic analysis of survival in challenge tests of furunculosis and ISA in Atlantic salmon. Genetic parameter estimates and model comparisons. Aquaculture. 2007;**272**(Suppl. 1):S297-S298. DOI: 10.1016/j.aquaculture.2007.07.155

[67] Kjøglum S, Henryon M, Aasmundstad T, Korsgaard I. Selective breeding can increase resistance of Atlantic salmon to furunculosis, infectious salmon anaemia and infectious pancreatic necrosis. Aquaculture Research. 2008;**39**:498-505. DOI: 10.1111/j.1365-2109.2008.01904.x

[68] Gjøen HM, Refstie T, Ulla O, Gjerde B. Genetic correlations between survival of Atlantic salmon in challenge and field tests. Aquaculture. 1997;**158**:277-288. DOI: 10.1016/S0044-8486(97)00203-2

[69] Ødegård J, Olesen I, Gjerde B, Klemetsdal G. Positive genetic correlation between resistance to bacterial (furunculosis) and viral (infectious salmon anaemia) diseases in farmed Atlantic salmon (*Salmo salar*). Aquaculture. 2007;**271**:173-177. DOI: 10.1016/j.aquaculture.2007.06.006

[70] Perry GML, Tarte P, Croisetière S, Belhumeur P, Bernatchez L. Genetic variance and covariance for 0+ brook charr (*Salvelinus fontinalis*) weight and survival time of furunculosis (*Aeromonas salmonicida*) exposure. Aquaculture Research. 2004;**235**:263-271. DOI: 10.1016/j.aquaculture.2004.03.002

[71] Henryon M, Berg P, Olesen NJ, Kjær TE, Slierendrecht WJ, Jokumsen A, et al. Selective breeding provides an approach to increase resistance of rainbow trout (*Oncorhynchus mykiss*) to the diseases, enteric red mouth disease, rainbow trout fry syndrome, and viral haemorrhagic septicaemia. Aquaculture. 2005;**250**:621-636. DOI: 10.1016/j.aquaculture.2004.12.022

[72] Silverstein JT, Vallejo RL, Palti Y, Leeds TD, Rexroad CE, Welch TJ, et al. Rainbow trout resistance to bacterial cold-water disease is moderately heritable and is not adversely correlated with growth. Journal of Animal Science.

2009;**87**:860-867. DOI: 10.2527/jas.2008-1157

[73] Hollebecq MG, Faivre B, Bourmaud C, Michel C. Spontaneous bactericidal and complement activities in serum of rainbow trout (*Oncorhynchus mykiss*) genetically selected for resistance or susceptibility to furunculosis. Fish and Shellfish Immunology. 1995;**5**:407-426. DOI: 10.1006/fsim.1995.0040

[74] Henryon M, Jokumsen A, Berg P, Lund I, Pedersen P, Olesen NJ, et al. Genetic variation for growth rate, feed conversion efficiency, and disease resistance exists within a farmed population of rainbow trout. Aquaculture. 2002;**209**:59-76. DOI: 10.1016/S0044-8486(01)00729-3

[75] Houston RD, Taggart JB, Cézard T, Bekaert M, Lowe NR, Downing A, et al. Development and validation of a high-density SNP genotyping array for Atlantic salmon (*Salmo salar*). BMC Genomics. 2014;**15**:90. DOI: 10.1186/1471-2164-15-90

[76] Lien S, Gidskehaug L, Moen T, Hayes BJ, Berg PR, Davidson WS, et al. A dense SNP-based linkage map for Atlantic salmon (*Salmo salar*) reveals extended chromosome homeologies and striking differences in sex-specific recombination patterns. BMC Genomics. 2011;**12**:615. DOI: 10.1186/1471-2164-12-615

[77] Rothschild MF, Soller M. Candidate gene analysis to detect traits of economic importance in domestic livestock. PRO. 1997;**8**:13-22

[78] Yamaguchi T, Dijkstra JM. Major histocompatibility complex (MHC) genes and disease resistance in fish. Cell. 2019;**8**:378. DOI: 10.3390/cells8040378

[79] Miller KM, Winton JR, Schulze AD, Purcell MK, Ming TJ. Major histocompatibility complex loci are associated with susceptibility of Atlantic salmon to infectious hematopoietic necrosis virus. Environmental Biology of Fishes. 2004;**69**:307-316. DOI: 10.1023/B:EBFI.0000022874.48341

[80] Langefors A, Schantz VT, Widegren B. Allelic variation of MHC class II in Atlantic salmon: A population genetic analysis. Heredity. 1998;**80**:568-575. DOI: 10.1046/j.1365-2540.1998.00321.x

[81] Langefors A, Lohm J, Grahn M, Andersen O, Schantz TV. Association between major histocompatibility complex class IIB alleles and resistance to *Aeromonas salmonicida* in Atlantic salmon. Proceedings of the Royal Society B: Biological Sciences. 2001;**268**:479-485. DOI: 10.1098/rspb.2000.1378

[82] Kjøglum S, Larsen S, Bakke HG, Grimholt U. How specific MHC class I and class II combinations affect disease resistance against infectious salmon anaemia in Atlantic salmon (*Salmo salar*). Fish and Shellfish Immunology. 2006;**21**:431-441. DOI: 10.1016/j.fsi.2006.02.001

[83] Gelderman H. Investigations on inheritance of quantitative characters in animals by gene markers I. Methods. Theoretical and Applied Genetics. 1975;**46**:319-330. DOI: 10.1007/BF00281673

[84] Davey JW, Hohenlohe PA, Etter PD, Boone JQ, Catchen JM, Blaxter ML. Genome-wide genetic marker discovery and genotyping using next-generation sequencing. Nature Reviews Genetics. 2011;**12**:499-510. DOI: 10.1038/nrg3012

[85] Palti Y, Gao G, Liu S, Kent MP, Lien S, Miller MR, et al. The development and characterization of a 57K SNP array for rainbow trout. Molecular Ecology Resources.

2014;**15**:662-672. DOI: 10.1111/1755-0998.12337

[86] Houston RD, Davey JW, Bishop SC, Lowe NR, Mota-Velasco JC, Hamilton A, et al. Characterisation of QTL-linked and genome-wide restriction site-associated DNA (RAD) markers in farmed Atlantic salmon. BMC Genomics. 2012;**13**:244-244. DOI: 10.1186/1471-2164-13-244

[87] Everett MV, Miller MR, Seeb JE. Meiotic maps of sockeye salmon derived from massively parallel DNA sequencing. BMC Genomics. 2012;**13**:521. DOI: 10.1186/1471-2164-13-521

[88] Caputo V, Giovannotti M, Nisi Cerioni P, Splendiani A, Olmo E. Chromosomal study of native and hatchery trouts from Italy (*Salmo trutta* complex, Salmonidae): Conventional and FISH analysis. Cytogenetics and Genome Research. 2009;**124**:51-62. DOI: 10.1159/000200088

[89] Guyomard R, Boussaha M, Krieg F, Hervet C, Quillet E. A synthetic rainbow trout linkage map provides new insights in to the salmonid whole genome duplication and the conservation of synteny among teleosts. BMC Genetics. 2012;**13**:15. Available from: http://www.biomedcentral.com/1471-2156/13/15

[90] Gharbi K, Gautier A, Danzmann RG, Gharbi S, Sakamoto T, Hoyheim B, et al. A linkage map for brown trout (*Salmo trutta*): Chromosome homeologies and comparative genome organization with other salmonid fish. Genetics. 2006;**172**:2405-2419. DOI: 10.1534/genetics.105.048330

[91] Woram RA, McGowan C, Stout JA, Gharbi K, Ferguson MM, Hoyheim B, et al. A genetic linkage map for Arctic char (*Salvelinus alpinus*): Evidence for higher recombination rates and segregation distortion in hybrid versus pure strain mapping parents. Genome. 2004;**47**:304-315. DOI: 10.1139/g03-127

[92] Ozaki A, Sakamoto T, Khoo S, Nakamura K, Coimbra MR, Akutsu T, et al. Quantitative trait loci (QTLs) associated with resistance/susceptibility to infectious pancreatic necrosis virus (IPNV) in rainbow trout (*Oncorhynchus mykiss*). Molecular Genetics and Genomics. 2001;**265**:23-31. DOI: 10.1007/s004380000392

[93] Baerwald MR, Petersen JL, Hedrick RP, Schisler GJ, May BA. Major effect quantitative trait locus for whirling disease resistance identified in rainbow trout (*Oncorhynchus mykiss*). Heredity (Edinb.). 2010;**106**:920-926. DOI: 10.1038/hdy.2010.137

[94] Davidson WS, Koop BF, Jones SJM, Iturra P, Vidal R, Maas A, et al. Sequencing the genome of the Atlantic salmon, *Salmo salar*. Genome Biology. 2006;**11**:403. Available from: http://genomebiology.com/2010/11/9/403

[95] Zhu Z, Chung WH, Shim EY, Lee SE, Ira G. Sgs1 helicase and two nucleases Dna2 and Exo 1 resect DNA double strand break ends. Cell. 2008;**134**(6):981-994. DOI: 10.1016/j.cell.2008.08.037

[96] Palti Y, Parsons JE, Thorgaard GH. Identification of candidate DNA markers associated with IHN virus resistance in backcrosses of rainbow (*Oncorhynchus mykiss*) and cutthroat trout (*O. clarki*). Aquaculture. 1999;**173**:81-94. DOI: 10.1016/S0044-8486(98)00471-2

[97] Moen T, Fjalestad KT, Munck H, Gomez-Raya L. A multistage testing strategy for detection of quantitative trait Loci affecting disease resistance in Atlantic salmon. Genetics. 2004;**167**:851-858. DOI: 10.1534/genetics.103.013227

[98] Moen T, Sonesson AK, Hayes B, Lien S, Munck H, Meuwissen TH. Mapping of a quantitative trait locus for resistance against infectious salmon anaemia in Atlantic salmon (*Salmo salar*): Comparing survival analysis with analysis on affected/resistant data. BMC Genetics. 2007;**8**:53. DOI: 10.1186/1471-2156-8-53

[99] Robledo D, Taggart JB, Ireland JH, McAndrew BJ, Starkey WG, Haley CS, et al. Gene expression comparison of resistant and susceptible Atlantic salmon fry challenged with infectious pancreatic necrosis virus reveals a marked contrast in immune response. BMC Genomics. 2016;**17**:279. DOI: 10.1186/s12864-016-2600-y

[100] Schmeisser H, Mejido J, Balinsky CA, Morrow AN, Clark CR, Zhao T, et al. Identification of alpha interferon-induced genes associated with antiviral activity in Daudi cells and characterization of IFIT3 as a novel antiviral gene. Journal of Virology. 2010;**84**:10671-10680. DOI: 10.1128/JVI.00818-10

[101] Gervais O, Barria A, Papadopoulou A, Gratacap RL, Hillestad B, Tinch AE, et al. Exploring genetic resistance to infectious salmon anaemia virus in Atlantic salmon by genome-wide association and RNA sequencing. BMC Genomics. 2021;**22**:345. DOI: 10.1186/s12864-021-07671-6

[102] Skugor S, Jorgensen SM, Gjerde B, Krasnov A. Hepatic gene expression profiling reveals protective responses in Atlantic salmon vaccinated against furunculosis. BMC Genomics. 2009;**10**:503. DOI: 10.1186/1471-2164-10-503

Chapter 5

Investigation of Trace Metal Bioaccumulation in Wastewater-Fed Fish: A Case Study

Aslihan Katip

Abstract

It was stated that the use of urban wastewater in food production in the 1970s and 1980s may lead to the development of alternative farming systems in the future. Fish fed with wastewater are grown in Asian countries. However, due to the mixing of domestic wastewater with industrial wastewater, many toxic micro-polluting wastewaters affect fish farming even more. The objectives of this study were to investigate the suitability of fish for human food consumption in terms of metals, to provide a basis for the development of a standard on the concentration of heavy metals in reclaimed water used for fish aquaculture, and to search the possibilities of technical improvement of the system in terms of more efficient wastewater treatment. This study will be useful in terms of precautions and disadvantages that can be taken against food shortages that may be experienced with the effect of climate change.

Keywords: bioaccumulation, fish, trace metals, transfer factor, wastewater

1. Introduction

It is estimated that one billion people depend upon freshwater fish as the prime source of protein [1]. Fish consumption makes a major contribution to nutrition, especially for the poorest (e.g., in Cambodia, Laos, and China). Therefore, it is useful to look briefly at the conclusions of the IPCC AR4 on fisheries. Fisheries will come under pressure from increased temperature stress and rising Ph associated with global warming. The frequency of extreme droughts and floods will have a disproportionate effect on fish habitat and populations, and the incidence of diseases is expected to rise. This will result in species extinctions at the margins of their current habitats (e.g., salmon and sturgeon), and fish yields in places like Lake Tanganyika are expected to fall by around 30 percent [2]. Cities will generate increasingly large amounts of effluent that will be recycled for agriculture, subject to water quality and health and safety considerations.

"Water reuse" refers to the production of water through water treatment processes, which introduces a feedback loop in the water cycle. Water reuse presents environmental, economic, and social benefits but also potential drawbacks. Treated wastewater was used in urban uses (green area irrigation, vehicle washing, fire extinguishing, urban pools and toilet water, etc.), industrial (cooling, boiler feeding, process water, etc.), agricultural irrigation, groundwater feeding, direct

IntechOpen

or indirect drinking water. Also, it could be used for feeding and improving surface waters and for fish production [3]. The reuse of wastewater for different purposes is even more important these days when there is a danger of drought [4].

1.1 Treatment and advanced treatment applications that could be used for wastewater feeding fishes

Point and diffuse pollutants are converted into end products such as CO_2, N_2, H_2S, and biomass by being mineralized (decomposed) by natural treatment processes (with the cooperation of bacteria/archae and algae) in rivers, wetlands, estuaries, and seas. A similar separation occurs in wastewater treatment plants [5]. Considering highly treated wastewater as a new water source will become more important in the future, as river flows are predicted to decrease by 20–30% due to global climate change [5]. After the water consumed as drinking/utility water is transformed into wastewater, it can be brought to suitable water quality for different reuse alternatives with the second, third, or advanced treatment stages [6].

Wastewater-fed fish culture has a history of more than a century in Germany. First, it receives well-treated wastewater from wastewater treatment systems. The latter is designed to treat raw wastewater that has been mechanically pretreated only. Net fish yield from wastewater-fed fish ponds is 500 kg/ha/7 months on average (estimated as 860 kg/ha/year), and loading rates are equal to 2000 persons/ha/day [7].

There are still serious psychological, social, and etic hesitations in front of the use of domestic wastewater with advanced treatment, even if it is brought to the quality of tap water, directly as drinking and utility water. It is known that water of this nature is given to aquifers and then drawn by wells and distributed to cities from a separate network (purple network) and used as B quality/class water at 50% lower cost for irrigation, WC flushing water, or industrial process water supply. The most courageous application in which treated wastewater of this quality was used as drinking water in pet bottles, called new water (NEWater), was made in Singapore [8]. The current legislation on the reuse of wastewater in Turkey was published in 2010. "Wastewater Treatment Plants Technical Procedures Communiqué" (Official Gazette no: 27527). In this communique, the selection of treatment technology, design criteria, and technical procedures for reuse of wastewater originating from settlements were given. According to the Communiqué, the main areas of use of treated wastewater were agriculture, industry, aquifer feeding, indirect firewater, use in toilets, and direct drinking water [9].

However, among the areas where treated wastewater can be reused, the most accepted ones are irrigation for agriculture and landscape purposes. The lowest accepted usage areas are direct use in the kitchen and bathroom [10]. Therefore, it is of great importance to increase the low rate of acceptance of the public in using this water, despite the reuse of wastewater by using appropriate engineering techniques [3].

In this case, as the dilution capacity of the streams will decrease, it may be necessary to apply "Ozone Oxidation + Granular Activated Carbon Filtration" at the Advanced Biological WWTP outlet [5]. In the removal of viruses, ultrafiltration membrane application and maturation pools and UV applications for the removal of other pathogenic microorganisms have been determined to provide the desired purification efficiency to a large extent [11]. Smin >3 (1 unit wastewater +2 units river/lake water) criterion can be taken as a measure for the minimum dilution in the discharge of low pollution (gray water or equivalent pollutant) used or treated wastewater into surface waters (streams and lakes). Absolute water should be more than two times of treated wastewater. It is thought that the domestic wastewater that has undergone advanced treatment and disinfection can be mixed with more than two times of clean water and used in the production of aquaculture.

In this study, trace metal concentrations in muscle, gill, and liver tissues of *Carassius gibelio* specie fed with wastewater from Bursa Water and Sewage Administration East Treatment Place were investigated. Their bioaccumulations and health risks (transfer factors—TF, bio-concentration factors—BCF, and hazard quotient—HQ) were computed and evaluated by comparison with metal concentrations in wastewater. This study was ensured useful and valuable information for evaluating potential health risks in wastewater recovery as aquaculture feeding water.

2. Materials and methods

2.1 Study locations

East Wastewater Treatment Plant treats the wastewater of the eastern part of Bursa City. It covers an area of approximately 250,000 m2 and wastewater of about 1,550,000 inhabitants is mixed with the facility. The 2017 flow rate of the treatment plant is 240,000 m^3/day. It is designed as 320,000 m^3/day for the year 2030. The Wastewater Treatment Plant is discharged into Deliçay Stream, which is a tributary of Nilüfer Stream, in the Susurluk River Basin. The Biological Treatment Plant is a five-stage Bardenphod that removes nitrogen and phosphorus [12].

2.2 Sample handling and analysis of water and fish tissues

The species *Carrassius gibelio* examined in this study has been recognized as an invasive species by the Republic of Turkey, and its prey has been released throughout the year [13]. The metal concentrations in the muscles, gills, and livers of fish fed with the effluent of the wastewater treatment plant were investigated seasonally. The investigated metals were chosen among the most common ones (Fe, Mn, Cu, Zn, Cr, Pb, Cd, Ni, As, and B) in wastewaters and fish.

Measurements were made by taking three fish samples (*Carrassius gibelio*) in each season in 2011–2012. The sizes of the fishes taken in polyethylene caps were measured in the laboratory. The tissues of muscle, liver, and gill were spared with stainless steel and homogenized. The tissue samples of 0.5 g (wet weight) in petri dishes were dried 24 hours in a drying oven. The samples in which dry weights were obtained were decomposed in a CEM Mars 5 Model microwave device by placing them in HP500 Teflon containers and adding 7 ml of nitric acid (HNO_3) and 1 ml of hydrogen peroxide (H_2O_2) [14]. After filtering, the water samples were acidified with 0.2% (v/v) nitric acid and stored in glass bottles [15]. Water and fish samples were taken and prepared simultaneously.

Trace elements in water and all fish tissues were measured with the ICP-OES device (VISTA-MPX model-VARIAN brand) [16].

2.3 Determination of metal bioaccumulations and risk assessment

Metal concentrations (based on wet and dry weight) in muscle, gill, and liver tissues were evaluated with national and international standards [17–22].

Transfer and bio-concentration factors (TF and BCF) were calculated to determine the level of bioaccumulation in fish tissues. Transfer factor was used to determine the amount of metal transferred from water or sediment to fish tissues [23]. TF and BCF formulations were given below [23, 24]:

$$TF = M_{tissue}\left(mg/kg\,dry\,weight\right)/M_{sediment\,or\,water}\left(mg/L\right) \qquad (1)$$

$$BCF = M_{tissue}\left(mg / kg \text{ wet weight}\right) / M_{water}\left(mg / L\right) \quad (2)$$

where M_{tissue} is the metal concentration in fish tissue; $M_{sediment}$, metal concentration in sediment. The concentrations in TF sediments were not used in this study because only the effect of water was examined.

BCF is calculated to see the effect of concentrations in water. BCF and TF are inversely proportional to exposure concentrations in the aquatic environment. In international references, it was stated that bioaccumulation was dangerous when BCF was >1000 and TF was >1. BCF and TF should be evaluated together to accurately determine the chronic effects [24, 25].

The consumption of *C. gibelio*, the fish species examined in this study, as the food was determined by the estimated daily intake (EDI) value [26, 27]:

$$EDI = \frac{Cfish * Dfish}{BW} \quad (3)$$

where C_{fish} = the average trace element concentration in fish muscle (μg/g dry weight), D_{fish} = the global average daily fish consumption (g/day) which was only 1.7 g/day for Turkey [28], and BW = average body weight (kg).

The USEPA was stated that the average body weight for an adult human for risk analysis was 70 kg [29]. The Hazard quotient (HQ) was calculated by dividing the estimated daily intake (EDI) by the established RfD (reference dose) to assess the health risk from fish consumption. It was stated that there was no significant risk when the HQ value was less than 1 [26].

3. Results and discussion

3.1 Muscle

Order of magnitude of the metal concentrations in muscle was as follows: Fe > Zn > B > Pb > Ni > Mn > Cu > Cr > Cd > As. It was determined that Mn, Cr, Pb, Cd, and Zn were higher and Cu, Ni, Fe, and As were lower than FAO and WHO standard values. Mn and Zn were determined in lower concentrations compared to Turkish and British standards. Metal concentrations determined in muscle, gill and liver tissues and national-international standard values are given in **Tables 1–3**, respectively.

3.2 Gill

Order of magnitude of the metal concentrations in gill tissue was as follows: Zn > Fe > Mn > B > Pb > Ni > Cu > Cr > Cd > As. Fe, Mn, Zn, Cr, Pb, Cd, and As were determined higher and Cu and Ni were lower than FAO and WHO standards. Mn was lower than Turkish standards. According to Turkish standards, other metals were evaluated similar to FAO/WHO standards.

3.3 Liver

Order of magnitude of the metal concentrations in liver tissue was as follows: Fe > Zn > B > Pb > Cu > Ni > Mn > Cr > Cd > As. Fe, Mn, Zn, Cr, Pb, and Cd were determined higher and Cu, Ni, As were lower than FAO and WHO standards.

Comparing the Turkish and English standards, Mn and Pb were determined as lower, and other metals were determined similar evaluating FAO/WHO standards.

Element (ww/dw)	Muscle (mg kg^{-1})	FAO, 1983/ WHO (dw mg kg^{-1})	Turkish Guidelines/ TFC (ww mg kg^{-1})	England (ww mg kg^{-1})
Fe	15.273 ± 7.30 / 82.686 ± 39.55	100		
Mn	0.855 ± 1.316 / 4.628 ± 7.123	1	20	
Cu	0.777 ± 0.563 / 4.205 ± 3.047	30	20	20
Zn	9.453 ± 3.102 / 51.169 ± 16.791	50/100	50	50
Cr	0.420 ± 0.399 / 2.273 ± 2.159	1		
Pb	1.046 ± 0.784 / 5.661 ± 4.243	0.5	1 / 0.3	2

Table 1.
Metals concentrations determined in muscle tissues and national-international standard values.

Elements (ww/dw)	Gill (mg kg^{-1})	FAO, 1983/WHO (dw mg kg-1)	Turkish Guidelines/ TFC (ww mg kg^{-1})	England (ww mg kg-1)
Fe	54.322 ± 20.051 / 287.309 ± 106.049	100		
Mn	4.088 ± 1.410 / 21.621 ± 7.404	1	20	
Cu	0.800 ± 0.350 / 4.231 ± 1.851	30	20	20
Zn	131.520 ± 45.916 / 695.609 ± 242.849	50/100	50	50
Cr	0.604 ± 0.377 / 3.194 ± 1.993	1		
Pb	1.481 ± 0.628 / 7.833 ± 3.321	0.5	1 / 0.3	2

Table 2.
Metal concentrations determined in gill tissues and national-international standard values.

Elements (ww/dw)	Liver (mg kg^{-1})	FAO, 1983/WHO (dw mg kg-1)	Turkish Guidelines/ TFC (ww mg kg^{-1})	England (ww mg kg-1)
Fe	202.25 ± 153.018 / 1104.576 ± 835.63	100		
Mn	0.961 ± 0.707 / 5.248 ± 3.860	1	20	
Cu	1.270 ± 0.683 / 6.935 ± 3.729	30	20	20
Zn	97.523 ± 65.213 / 532.573 ± 356.128	50/100	50	50
Cr	0.438 ± 0.139 / 2.391 ± 0.759	1		
Pb	1.613 ± 0.839 / 8.808 ± 4.581	0.5	1 / 0.3	2

Table 3.
Metals concentrations determined in liver tissues and national-international standard values.

Cd, Ni, As, and B elements determined in muscle, gill and liver tissues, and national-international standard values are given in **Table 4**.

The metal concentrations and accumulation amounts (g/day/body weight) obtained in this study could be used to form a guide value for metal intake. Similar

Element (ww/dw)	Tissue	Concentration (mg kg^{-1})	FAO, 1983/WHO (dw mg kg^{-1})	Turkish Guidelines/ TFC (ww mg kg^{-1})	England (ww mg kg^{-1})
	Muscle	0.229 ± 0.264 / 1.239 ± 1.429	0.5/1	0.1 / 0.05	0.2
Cd	Gill	0.248 ± 0.306 / 1.311 ± 1.618			
	Liver	0.289 ± 0.275 / 1.578 ± 1.501			
	Muscle	0.966 ± 0.945 / 5.228 ± 5.115	10		
Ni	Gill	1.004 ± 0.694 / 5.310 ± 3.670			
	Liver	0.977 ± 0.645 / 5.335 ± 3.522			
	Muscle	0.042 ± 0.0236 / 0.227 ± 0.1277	0.27		
As	Gill	0.0585 ± 0.0267 / 0.309 ± 0.141			
	Liver	0.0443 ± 0.0296 / 0.241 ± 0.161			
B	Muscle	1.575 ± 1.457 / 8.525 ± 7.886			

Table 4.
Cd, Ni, As, and B elements determined in muscle, gill and liver tissues and national-international standard values.

studies should be done with different fish species [27]. The effects of heavy metals on alive changes depending on their concentrations, type of organism, ionic properties of metals (solubility value, chemical structure, ability to form redox and complexes) , tissue in which they are taken into the body and way of intake. Also, other minerals in the ambience and chemical properties of water effect the metal bioaccumulations. Because of these reasons, the physicochemical properties of the water used in aquaculture should be investigated and limited by legal values [30].

3.4 Metal bioaccumulations and risk assessment

The treated feed wastewater (TFE) was improved with national and international standard values. It was determined that most of the metals examined were above portable water standards [31–33] and USEPA surface water standards for toxic commentating [34]. There is no standard value for Zn and Cu in the Turkish Fisheries Regulation [35]. However, all metals were below the "Irrigation Water of Technical Methods Notification of Wastewater Treatment Plants" [36].

Except for Cd, all the other metals were found below the standard values of the People's Republic of China Fisheries Regulation-GB 8978 [37]. In light of these evaluations, it was been determined that the wastewater fed by the fishes was suitable for irrigation standards but not suitable for some parameters for aquaculture.

Annual and seasonal averages of transfer factors (TF), bio-concentration factors (BCF), and estimated daily intake values (EDI) were computed by using metal concentrations in treated wastewater effluent and examined fish tissues. The calculated factors and values provided a better assessment of the accumulation levels of metals in fish and the health risks that may occur when consumed by humans as

Metals	Treated effluent concentrations (mg L^{-1})	Muscle		Gill		Liver		US EPA (1999)	EDI (μg/kg b.w/day)	RfD μg/kg b.w/ day USEPA, 2005	Hazard quotient (EDI/RfD)
		TF(L/kg)	BCF(L/kg)	TF(L/kg)	BCF(L/kg)	TF(L/kg)	BCF(L/kg)	BCF (L/kg ww)			
Fe	0.4254 ± 0.2935	194.372	35.903	675.386	127.696	2596.560	475.456	—	2.008	700	0.0028
Mn	0.0988 ± 0.0513	47.224	8.724	220.622	41.715	53.551	9.806	—	0.112	140	0.0008
Cu	0.0282 ± 0.0247	149.113	27.553	150.035	28.370	245.922	45.035	710	0.102	40	0.0025
Zn	0.1270 ± 0.0731	402.306	74.433	5477.236	1035.590	4193.490	767.900	2059	1.242	300	0.0004
Cr	0.035 ± 0.0224	64.943	12.514	91.257	17.257	68.314	12.514	19	0.055	3	0.0183
Pb	0.0379 ± 0.0206	149.367	27.599	206.675	39.076	232.401	42.560	0.09	0.137	0.05	2.74
Cd	0.0153 ± 0.0168	80.392	14.967	85.686	16.209	103.137	18.890	907	0.030	1	0.030
Ni	0.0403 ± 0.0229	129.727	23.970	131.762	24.913	132.382	24.243	78	0.126	1.5	0.084
As	0.0033 ± 0.0011	68.788	12.727	93.636	17.727	73.030	13.424	114	0.005	0.3	0.0166
B	0.3393 ± 0.0970	25.125	4.642	24.721	4.674	61.930	11.340	—	0.207	—	

**EDI values were calculated for only muscle tissue due to human consumption [12].*

Table 5.
The annual averages of metal concentrations in treated effluent and calculated TF, BCF, EDI, HQ values.

food. The values of TF, BCF, EDI, HQ, and the annual average metal concentrations of treated effluent are presented in **Table 5**. Computed TF values of all metals in all tissues were determined above the 1. Except for Pb, all other elements were found to be lower than the USEPA BCF limit values. It is known that the TF value gives more realistic results than the BCF values. Large BCF values indicate low chronic effects and low potential for secondary poisoning. In other words, large BCF indicates that there is no high danger. No value can show the hazardous status for BCF values. BCF values of most metals (such as iron) are above 1000 in healthy aquatic ecosystems. Metals have a greater BCF value in systems without contamination. Transfer factors were evaluated since the possibility of coronal effect and the danger status could not be evaluated with BCF [23, 25]. The TF values of all metals examined in this study were found to be above 1. This value shows that metals could bioaccumulate and had potential health effects.

Annual and seasonal averages of TF and BCF values in fish tissues showed that Zn and Fe were high and B and Mn were low values. The order of the annual mean TF and BCF values calculated in the tissues was the same. It was found as Zn > Fe > Pb > Cu > Ni > Cd > As>Cr > Mn > B in muscle, as Zn > Fe > Mn > Pb > Cu > Ni > As>Cr > Cd > B in gill, and as Zn > Fe > Cu > Pb > Ni > Cd > As>Cr > B > Mn in liver. Fe, Zn, and Cu were found to be higher according to the seasonal means of TF and BCF values for the three tissues. Similarities were found in seasonal changes in tissues. According to the calculations for both factors, Cr, Pb, Ni, and B values in muscle were determined as higher in the summer season, Cd was raised in spring, Mn was raised in autumn, and Zn was raised in winter.

Nevertheless, seasonal differences of As, Fe, and Cu elements were found for both factors. BCF values of As, Fe, and Cu were higher in autumn, TF values of As and Fe were higher in summer months, and TF value of Cu was higher in spring. For both factors, Cr, Cd, and Zn values in gill tissues were found higher in spring, Ni and Fe were higher in summer, and Cu was found higher in winter. However, while BCF values of As, Pb, B, and Mn were higher in autumn, TF values of As, B, and Mn were higher in summer months, and TF value of Pb was higher in spring. For both factors, As, Cd, Mn, and Zn values in liver tissues were found higher in autumn, Cr, Cu, and Fe were higher in winter, and Pb and Ni were found higher in spring. However, TF value of B element was higher in summer; BCF value of B was higher in autumn. The annual means of metal concentrations in the tissues were found to differ in the order of magnitude, but according to FAO and WHO standards, the same elements were found to be high (Mn, Zn, Cr, Cd, and Pb) and low (Cu and Ni) in all three tissues. Ni concentrations was over than *C. gibelio* species exist in other water resources. Also, Zn, Cr, and Pb were over than the different fish types. There were differences between the order of magnitude of the concentrations in tissues and the order of magnitude of TF and BCF. While B and Mn concentrations were high in all tissues, the order of bioaccumulation factors of these elements was lower than other elements. Also, the concentrations of Cd and As were lower than other elements; however, their bioaccumulation factors were found higher than the others in all tissues.

It was determined that all elements bioaccumulated in the three tissues according to TF values. TF and BCF values of Fe, Zn, and Cu elements had the highest values in all tissues. The metal concentrations in summer and autumn were higher than in the other seasons. Nevertheless, seasonal differences of bioaccumulation factors were determined distinct from concentration alteration.

The element concentrations apart from B and Fe were determined higher in all tissues in summer, and TF and BCF calculations were determined higher in different seasons. Metal concentrations other than As and B in effluent water were higher in summer and autumn than in the other seasons like concentrations in fish. Nevertheless, the correlations among the Cd, Mn, Pb, and Cu concentrations in all tissues and effluent water were determined statistically important. The correlations calculated for Cd, Mn, Pb, and Cu elements were found to be significant, indicating that the bioaccumulation was due to effluent. Seasonal changes of other elements' biological accumulation factors and their concentrations in the effluent were found different. Due to these reasons, it was considered that baits and sediment layer of the feeding pool could affect the bioaccumulations in the fishes.

According to EDI and HQ values (**Table 5**), it was observed that there is only a carcinogenic risk in terms of Pb among all metals. Finding the HQ value of Pb greater than 1 indicates a carcinogenic risk. In addition, lead prevents the enzyme systems from working because it imitates the metabolic behavior of the calcium element. Pb is toxic and causes brain damage [29].

4. Conclusions

The results of these studies showed that the treated wastewater used in fish feeding is suitable for irrigation water, but not for aquaculture. Metal concentrations in the fish tissues were determined as over than the standards. The concentrations of liver and gill were higher than muscle. It was determined that investigated metals (Fe, Mn, Cu, Ni, Zn, Cr, Pb, Cd, As, and B) were bioaccumulated in all tissues. HQ values of Pb element in muscle tissue had carcinogenic risk and BCF value of only Pb among all elements was higher than limit values. It was determined that Pb and Cd, which were the most hazardous metals, were higher than the international regulations. For this reason, it has been determined that the examined fish were not suitable for human and animal edible.

Suggestions for the improvement of this study were presented below:

- In the future, the effects of sediment layer and feed on metal accumulation in fish fed in wastewater-fed ponds should be investigated.

- The effects of different pollutants on different wastewater-fed fish species should be investigated.

- These studies could be used in future studies to establish a guide value for metal accumulation that should be taken per day according to body weight (g/day/body weight).

- Re-use of water in the industry must be done. Thus, an approach compatible with the circular economy approach is followed. There might be health and ethical risks in its use for food production. To feed aquatic products in wastewater, projects that include advanced toxicology experiments should be carried out to reveal the toxic effects that may occur in the long term.

- In-depth research on primary pollutants (pharmaceutical residues, personal care products, industrial chemicals, endocrine disruptors, etc.) in wastewater was required. In addition to the benefits, the risks also need to be evaluated correctly. Considering all these issues, it was concluded that it would be more appropriate to work with advanced treated domestic sewage treatment water that does not mix with industrial water.

Author details

Aslihan Katip
Faculty of Engineering, Department of Environmental Engineering, Bursa Uludag University, Bursa, Turkey

*Address all correspondence to: aballi@uludag.edu.tr

References

[1] IRD. 2010. Tropical Marine Ecosystem Programme. Available from: http://www.mpl.ird.fr/suds-en-ligne/ecosys/ang_ecosys/intro1.htm

[2] FAO, 2008. FAO Water Reports: Climate Chance, Water and Food Security

[3] Katip A. Evaluation of treated wastewater reuse areas. Ömer Halisdemir University Journal of Engineering Sciences. 2018;7(2):541-557

[4] Solak Z. Urban Treated Waste Water Reuse (Hendek Example). Master Thesis, Sakarya University Institute of Science; 2018.

[5] Fakıoğlu M, Güven H, Öztürk İ. Reuse of treated wastewaters: Scientific facts, psychological and religious hesitations. Journal of Halal & Ethical Research. 2020;**2**(2):1-20

[6] Nas B, Yılmaz C. Benefits and risks in using treated domestic/urban wastewater as a new water source. Climate Change and Environment. 2019;**4**(2):42-46

[7] Edwards P., Pullin, R.S.V. Wastewater-Fed Aquaculture Systems: Status and Prospects. NAGA; 1996;**19**(1):33

[8] PUB NEWater, Singapore Public Utilities Board 2020. Available from: https://www.pub.gov.sg/watersupply/fournationaltaps/newater[Accessed: 17-05-2020]

[9] WTPTPC. Wastewater Treatment Plants Technical Procedures Communiqué, 20 March 2010, Official Gazette No: 27527

[10] Massoud MA, Kazarian A, Alameddine I, Al-Hindi M. Factors influencing the reuse of reclaimed water as a management option to augment water supplies. Environmental Monitoring Assessment. 2018;**190**(9):531

[11] Adalı S, Yalılı KM. Use of Treated Wastewater in Agricultural Irrigation: The Example of İznik. International Journal of Biosystems Engineering. 2020;**1**(1):12-23

[12] Katip A. Bioaccumulation of Trace Metals in Wastewater-Fed Aquaculture: A Case Study in Turkey Pol. J. Environ. Stud. 2019;**28**(6):4221-4238

[13] Tarkan AS. Yabancı Tatlısu Balıklarının Dünyada ve Türkiye'de Giriş Yolları, Etkileri ve Bunlardan Korunma Yöntemleri. Journal of Fisheries & Aquatic Sciences. 2013;**28**:63-104

[14] Uysal K, Köse E, Bülbül M, Dönmez M, Erdoğan Y, Koyun M, et al. The comparison of heavy metal accumulation ratios of some fish species in Enne Dame Lake (Kütahya/Turkey). Environmental Monitoring and Assessment. 2009;**157**:355-362

[15] Benzer S, Arslan H, Uzel N, Gül A, Yilmaz M. Concentrations of metals in water, sediment and tissues of Cyprinus carpio L., 1758 from Mogan Lake (Turkey), Iranian Journal of Fisheries. Sciences. 2013;**12**(1):45-55

[16] APHA. AWWA Standard Methods for the Examination of Water and Wastewater. 23rd ed. Washington DC USA: American Public Heallth Association; 2017

[17] FAO. Food and Agriculture Organization. Compilation of legal limits for hazardous substances in fish and fishery products. FAO Fishery Circular No. 464, pp. 5-100, 1983

[18] FAO/WHO. Evaluation of certain food additives and the contaminants

mercury, lead and cadmium. WHO Technical Report, Series No.505.1989

[19] El-Moselhy KM, Othman AI, El-Azem HA, Metwally MEA. Bioaccumulation of heavy metals in some tissues of fish in the Red Sea, Egypt. Egyptian Journal of Basic and Applied Sciences, I. 2014;**97-105**

[20] Akan JC, Mohmoud S, Yikala BS, Ogugbuaja VO. Bioaccumulation of Some Heavy Metals in Fish Samples from River Benue in Vinikilang, Adamawa State, Nigeria. American Journal of Analytical Chemistry. 2012;**3**:727-736

[21] Maff. Ministry of Agriculture, Fisheries and Food. Monitoring and surveillance of non-radioactive contaminants in the aquatic environment and activities regulating the disposal of wastes at sea, 1997. In: Aquatic Environment Monitoring Report No. 52. Lowestoft, UK: Center for Environment, Fisheries and Aquaculture Science; 2000

[22] TFC. Turkish Food Codes. Maximum limits of contaminants in foodstuffs. Official Gazette. 2011;**28157**

[23] Abdel-Baki AS, Dkhil MA, Al-Quraishy S. Bioaccumulation of some heavy metals in tilapia fish relevant to their concentration in water and sediment of Wadi Hanifah, Saudi Arabia. African Journal of Biotechnology. 2011;**10**(13):2541-2547

[24] USEPA. 1999b. Screening Level Ecological Risk Assessment Protocol for Hazardous Waste Combustion Facilities. United States Environmental Protection Agency EPA530-D-99-001C.1999

[25] OECD. Bioaccumulation of Metal Substances by Aquatic Organisms Part 1. OECD Meeting, Paris, September 7-8, 2011

[26] Onsaint S, Ke C, Wang KJ, Wang WX. Trace elements in two marine fish cultured in fish cultured in fish cages in Fujian province. China. Environ. Pollut. 2010;**15**:1334-1345

[27] Ekeanyanwu RC, Nwokedi CL, Noah UT. Monitoring of metals in Tilapia Nilotica tissues, bottom sediments and water from Nworie River and Oguta Lake in Imo State, Nigeria. African Journal of Environmental Science and Technology. 2015;**9**(8):682-690

[28] Sariözkan S. Türkiye'de Balıkçılık Sektörü ve Ekonomisi. Turkish Journal of Aquatic Sciences. 2016;**31**(1):15-22

[29] Mol S, Karakulak FS, Ulusoy S. Assessment of potential health risks of heavy metals to the general public in Turkey via consumption of red mullet, whiting, turbot from the Southwest Black Sea. Turkish Journal of Fisheries and Aquatic Sciences. 2017;**17**:1135-1143

[30] Jaishankar M, Tseten T, Anbalagan N, Mathew BB, Beeregowda KN. Toxicity, mechanism and health effects of some heavy metals. Interdisciplinary Toxicology. 2014;**7**(2):60-72

[31] Guidelines for Drinking Water Quality. 4th ed. Inc. Addendum; Word Health Organization; 2017

[32] EU. The Drinking Water Directive Directive 2015/83/EC 1787/83/EC. 2015

[33] TSE. Turkish Standardization Institute TS266 Standards - Water for Human Consumption (Turkish standards) Rep: 25730. Ankara: Ministry of Health; 2005

[34] USEPA. National Recommended Water Quality Criteria–Correction EPA 4304T, 2009

[35] ANONYMOUS. Aquaculture Law and Regulation. Ankara: Ministry of Agriculture and Rural Affairs; 2002. pp. 63-78

[36] ANONYMOUS. Notification of wastewater treatment plant technical procedures - criteria for the use of wastewater as irrigation water. Official Newspaper. 2010;**27527**:2010

[37] ANONYMOUS. National Standard of the People's Republic of China Waste Water Discard Standards.GB 8978-1996, 1996

Chapter 6

Sexual Maturation in Farmed Atlantic Salmon (*Salmo salar*): A Review

Patricia Rivera, José Gallardo, Cristian Araneda and Anti Vasemägi

Abstract

The sexual maturation of Atlantic salmon *Salmo salar* is a multifactorial process in which fish acquire somatic characteristics to reproduce. In salmon farming has been described a high variability in the trait age at maturation derived from wild reproductive strategies. Early maturation is a phenotype that generates serious economic repercussions on both, sea cage and on land-based aquaculture systems. In view of the challenges of this problem for the global salmon farming industry, it is essential to thoroughly understand the influencing factors of early and late maturation to find efficient alternatives for managing the phenomenon. This review briefly describes sexual maturation in *S. salar*, its variability in cultures, and the factors influencing the maturation age trait at the physiological, genetic and environmental levels. The control of early maturity through changes to the natural photoperiod and through the use of genetic markers are discussed.

Keywords: sexual maturation, *Salmo salar*, multifactorial, reproduction, cultures, variability, photoperiod, physiological, genetic markers, GWAS

1. Introduction

Sexual maturation in *S. salar* is a complex and multifactorial process whose purpose is to acquire the somatic and behavioral conditions necessary to perform reproductive functions. In both wild and domesticated populations, there is variability in the age at maturation – a characteristic known as a life history trait – which can occur at an early or late stage in both males and females. These reproductive strategies were adopted by wild congeners to increase their reproductive success, perpetuate the species, and promote population sustainability.

Some salmon farms worldwide, including large industries in countries such as Canada [1], land-based aquaculture systems [2], and others, have experienced significant economic losses associated with salmon presenting advanced maturation, either in the freshwater cycle or, with greater repercussions, in fattening stages prior to harvest. These losses occur because when maturation is reached, fish divert energy from body reserves to reproduction, resulting in salmon with a smaller body size and lower organoleptic quality of the fillet. During meat evaluation in processing plants, any degraded physical characteristic will result in a decrease in the commercial value of the fish.

The time and degree of maturation in *S. salar* are influenced by several factors, which can be intrinsic, such as genetic makeup, body composition and metabolic status, or environmental, and these factors can affect anatomical and physiological processes. Among environmental factors, photoperiod is considered a major determinant of the maturation of most cultured teleosts [3], and temperature influences the variation in age and size at maturity in salmonids [4].

Several strategies have been used by the industry to control early maturation in salmon cultures. Among these, photoperiod manipulation has been an alternative that has provided better results in terms of reducing the advanced-maturation phenomenon. Currently, thanks to knowledge regarding the genetic component of sexual maturation, new approaches are being investigated, such as molecular marker-assisted selection. Knowing that the goal of salmon farmers is to maintain salmon in an immature stage to preserve the quality of the product and ensure sustainable production, this review presents a summary of the following topics: life cycle and sexual maturation in *S. salar*; maturation variability in cultures; and physiological, genetic and environmental factors influencing maturation.

2. Life cycle of Atlantic salmon

Atlantic salmon is an anadromous species whose life cycle (**Figure 1**) in the wild can include a reproduction and rearing phase in freshwater and a growth and sexual maturation initiation phase in the ocean [5, 6]. In this general pattern, there are some alternative strategies, developed mainly by males, that can be very successful in the natural environment, such as, 1) sexual maturation in freshwater as precocious male parr; 2) sexual maturation of fish known as jacks, which mature prematurely in freshwater prior to sea transfer or which reach a body size of approximately 0.5 kg in the sea before maturation [7]; and 3) maturation of fish known as grilse, which reach maturity, with a body size of typically 2 to 5 kg, after 1.5 years in the sea [8] cited in [9]. In all the cases described above, reproduction and spawning occur in the autumn so that the eggs incubate in the gravel substrate during the winter and hatch after two or three months depending on the water

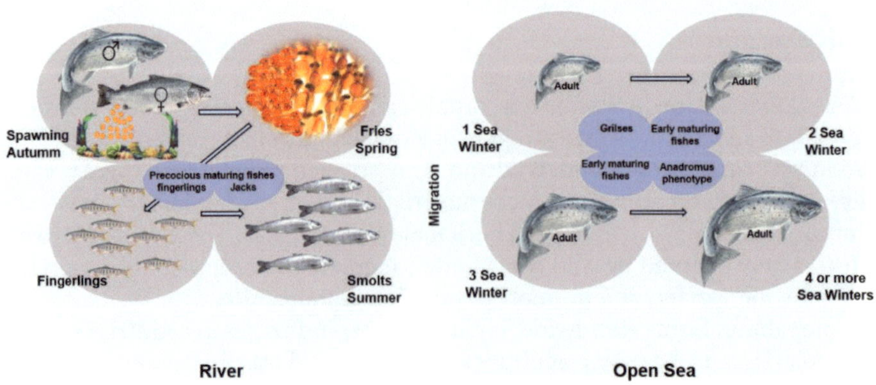

Figure 1.
Life cycle of Atlantic salmon. Both males and females of Atlantic Salmon, in the spawning season in autumn, fertilize and incubate eggs which, in spring hatch and will become fries. After a few months, the fries become to be fingerlings, and one to five years after hatching S. salar *juveniles typically descend as smolts to the ocean to feed and, return to the rivers of origin to a new cycle as adults. Early maturation could be presented in freshwater (precocious male parr); in freshwater prior to sea transfer (jacks), in the first sea winter after transfer to sea (grilse), and in a second or third winter (early maturation). While those adult salmon that mature after 4 or more years in the sea are known as anadromous or late maturing fish.*

temperature [10], so in some environments this process could be delay by the low temperatures of the water. One to five years after hatching, *S. salar* juveniles typically descend as smolts to the ocean to feed and, return with a high degree of fidelity to the rivers of origin to spawn and start a new cycle [11].

3. Maturation in domesticated populations of *S. salar* and its implications in production

The presence of different reproductive and maturation strategies among individuals within a wild population of Atlantic salmon has shown to reflect its genetic diversity [12]. That is, to maintain the natural sustainability of the species, sexual maturation in salmon occurs early or late in different proportions of individuals in both the freshwater and ocean environments [13]. This variability is common and is part of the life history of Atlantic salmon to perpetuate the species and maintain genetic diversity.

Variation in the maturation phenotypes beneficial to population survival in wild populations of Atlantic salmon can be undesirable in domesticated populations. For example, early maturation, especially in fattening stages prior to harvest, affects the profitability of commercial production because it is associated with the expression of secondary sexual characteristics [14]. Losses have also been observed due to the use of body reserves for gamete formation [9] and to decreased growth and degradation of fish meat [15]. These circumstances together hinder not only the profitability of fish farms but also the management, care, and survival of animals [14].

The strong negative impacts caused by the presence of high proportion of early maturing animals are especially noticable in large-scale commercial operations. [2] describe a serious production disaster in a land-based aquaculture system in which 80% of male salmon matured earlier than the harvest time, with recorded weights of 4 to 5 kg, forcing the fish farmers to close their facilities. In economic terms, [1] estimated significant losses of up 4,4% which represented between 11 to $24 million in 2002 in the salmon industry of New Brunswick, Canada, due to this phenomenon. As noted in both cases, early maturation often has irreparable economic consequences.

The incidence of the early maturation of individuals in domesticated populations is influenced by production goals because the industry seeks to obtain animals with a greater body weight in less time. [1] describe the main risk factors that cause early maturity in salmon; these factors include the intensity of feeding diets with a high fat content, the exaggerated manipulation of photoperiod regimes to delay spawning, and differences in water temperature, among others. As noted, the management of cultures in controlled environments contributes to the risk of early maturation, which is why fish farms find it necessary to resort to other control alternatives.

4. Physiology and factors that influence maturation in Atlantic salmon

Maturation is a process intrinsically governed by genetic makeup, body composition and metabolic status and by environmental factors [16]. Animals exhibit biological rhythms that are usually in step with environmental factors, which are also known as proximate and ultimate causes or factors [17]. Proximate factors affect anatomical and physiological processes, and ultimate factors underlie adaptation and diversification [18]. Therefore, integration of the totality of these signals in the brain is the main trigger for maturation and reproduction.

Of the environmental factors that can be categorized as proximate, photoperiod is the most important and is considered the main determinant of the maturation of most cultured teleosts [3], it has effects such as delayed onset of sexual maturity in salmonids [16] and increased maturational steroids and gonadal development [19]. In contrast, the temperature influences the variation in the age and size at maturity in salmonids [4] and can alter the time of ovulation [20]. Finally, inter and intra-specific competition maximize reproductive success within the limitations imposed by the opposite sex, the environment and phylogeny [21]. Together, all these factors modulate the maturational and reproductive patterns of salmon. The integration of all these signals when processes in the brain of the fish, especially seasonal changes in the photoperiod, triggers the beginning of maturation and modulates the seasonal patterns of endocrinological activity regulating the cyclic process, whose purpose is to achieve optimal reproduction. Notably, reproductive events occur at times during which salmon determine the survival of their progeny by ensuring the hatching of their larvae in seasons with abundant food resources [22].

In salmonids, the pineal gland and the brain play important roles in the activity of the reproductive axis because they develop light-perceptive, integrative and executive functions [23]. Light perception in these fish occurs through their photoreceptor organs, which transmit an electrical message at night that results from the release of melatonin [24]. The directive derived from this is the secretion of gonadotropin-releasing hormone (GnRH) by hypothalamic neurons and, subsequently, the secretion of gonadotrophins, follicle-stimulating hormone (FSH) and luteinizing hormone (LH), which regulate gametogenesis and gonadal steroidogenesis [3].

The intrinsic factors acting on the physiology of maturation include genetic component. Gene expression of the *vgll3* and variation together with other genomic regions associated with the hypothalamic pathway are analyzed in [25–27]. Based on the results from these studies, it could be suggested the aspects conducive to the maturation process, beginning with the expression of the *ywhab* gene, which contributes to the cytoplasmic retention of the *yap/taz* gene complex, which in turn acts as a cofactor for the expression of the *tead3* gene, ultimately interacting with the *vgll3* gene for the translation of the Vgll3 protein. The vgll3 gene has two alleles; the early maturity allele is expressed in Sertoli cells in males and in granulosa cells in females and acts in cell proliferation processes [27]. However, further research is still needed to understand the biological functions attributed to the genomic regions associated with maturation.

In summary, in a physiological context, sexual maturation can be understood as the process in which an animal begins the development of reproductive competencies through the integration of intrinsic and extrinsic factors. The maturational process is accompanied by a cascade of hormone secretion that modulates gonadal function in specific stages prior to reproduction, where the action of FSH occurs in early stages of gametogenesis, promoting the synthesis of sex steroids for spermatogonial proliferation in males and the progression of vitellogenesis in females and, subsequently, LH acts in the formation of maturational steroids [3].

5. Control of maturation in farmed Atlantic salmon

During fattening at sea, salmon must be maintained in an immature stage to preserve the quality of the product and ensure a good selling price. Therefore, various strategies have been implemented that help mitigate sexual maturation. For example, a common practice, but only effective in the short term, is eliminating or discarding early maturing males prior to entering the sea. Other alternatives are, for example, those described by [16], who cite, the induction of polyploidy mainly triploidy, modification of genes and the manipulation of environmental factors, especially photoperiod and temperature as [19]. Finally, recent findings regarding genomic regions that govern sexual maturation in Atlantic salmon [28] have shed light on the genetic determinants of this trait. Next, the use of photoperiod and genetic markers for the control of early sexual maturation is discussed in more detail.

5.1 Control of early maturation by photoperiod in marine cages and land-based aquaculture systems

Atlantic salmon marine cage aquaculture centers work with photoperiod protocols that seek to reduce the fish's perception of seasonal changes in photoperiod, mainly the decrease in daylight hours from summer to winter, thus inhibiting maturation. Examples of reductions in early maturation in cultures with different photoperiod regimes include the study by [29], who work with one-year-old Atlantic salmon smolts in the northern hemisphere treated with three different photoperiod regimes between January to July, with what, the proportion of sexually maturing fish was significantly lower among both sexes, from 91 to 9% in females and from 74 to 16% in males, in the groups treated from January with natural light + continuous additional light which was not switched off during daytime and was supplied by a 1300-W quartz halogen light. Additionally, in a study by [15], similar results were obtained, with 50% maturation in salmon reared under natural light and only 0.8% in males treated with artificial light for 9 months. More recently, the effectiveness of different alternative sources of artificial light has also been investigated. For example, [30] investigated the incidence of early sexual maturation in a culture of *S. salar* subjected to continuous artificial light, including five different LED light intensities, a single light intensity from a metal halide source and a control treatment with natural light; the results indicated that sexual maturation in males (6.1% under natural light) was arrested uniformly at all intensities. With this, the usefulness of this alternative light source on the control of maturation was confirmed, and widely used in fattening at sea.

The advancement of land-based aquaculture technologies has allowed salmon to complete the life cycle fully in these systems, but early maturation has been reported as a significant problem [19]. To address this issue, several studies have evaluated the efficacy of photoperiod manipulation to control early sexual maturation in these systems. For example, [31] showed that compared with Atlantic salmon exposed to continuous light for 24 hours during the first year, Atlantic salmon exposed to a reduced photoperiod of 18 hours of light from the first feeding to 1 year after hatching in a freshwater recirculation aquaculture system (RAS) showed a significantly higher proportion of mature males despite regimes of photoperiod, so fishes sampled at 19-months, were 50.0% grilse in the 18 h group and 33.3% grilse in the 24 h group. Additionally, [19] showed that the combination of spectral composition, photoperiod and light intensity can be used effectively to suppress or delay the gonadal development of Atlantic salmon reared in RAS. In summary, considerable number of studies demonstrate that photoperiod can be used effectively for the control of early maturation in both marine cages and land-based aquaculture systems.

5.2 Control of early maturation by genetic markers

Age of maturation in Atlantic salmon is a heritable trait so it should be possible to prevent early maturation using selective breeding [32–34]. Early records of high heritability on age at maturation (0.48 ± 0.20) and therefore great opportunity to control early maturation in Atlantic salmon were described from the mid-1980s [32]. High heritability values are usually expressed as large differences in the proportion of grilse between families of the same population. However, the systematic elimination of grilse in domesticated populations did not generate the expected results for reasons that will be explained later. This led, during the growth of the Atlantic salmon aquaculture in the 1990s, to the elimination of populations characterized as having high levels of grilse. More recent studies confirm that the age at sexual maturation [14] or some associated traits such as maturation in fresh water [6] or the proportion of grilse [35] have from medium to large heritable component and that there are significant differences between populations.

With the identification of the first genetic markers associated with the age of maturation trait in Atlantic salmon [14, 22], not only has a genetic explanation emerged for the variation previously described in early sexual maturation but foundations have also been laid to initiate more effective control of early maturation through the selection of fish that possess late-maturing genotypes (**Table 1**). The seminal studies of [14, 22], first allowed recognizing that there is a major/effect locus that controls the age of maturation, which has sex-dependent dominance and favors early maturation in males and late maturation in females [22]. This phenomenon may partially explain the great diversity of reproductive strategies observed in *S. salar*, particularly in male precocious parr, jacks and grilses. In addition, this phenomenon may also explain why the phenotypic elimination of early males does not have an impact on the control of early maturation in the long term, as heterozygous females can maintain the early maturation allele and pass it to their offspring. Those authors also identified that the locus with the largest effect on chromosome 25 is associated with the *vgll3* gene (vestigial-like family member 3 gene). This gene explained between 33 and 39% of the variation in age at maturation, an unexpectedly large proportion for a highly complex trait [14, 22].

In addition to abovementioned works, several genome-wide association studies (GWAS) and whole-genome sequencing studies (WGS) have identified additional genetic variants associated with sexual maturation in different wild and domesticated salmon populations. For example, [35], identified another quantitative trait locus (QTL) associated with the age of maturation in North American wild salmon populations; this marker is located on chromosome 6 and explained 6% of the variation in the early maturation phenotype. They also described that the frequency of the early genotype of the *vgll3* gene is lower in this population, contrary to that observed in European populations in which the early genotype is quite common. In turn, [28] reported contrasting results demonstrating that the *vgll3* gene was not associated with maturation in females in a domesticated population, known as Mowi, suggesting that other unidentified genomic regions control the age of maturation. More recently, based on the most extensive GWAS analysis in Atlantic salmon to date, [34] found significant associations with age of maturation in 28 of 29 chromosomes, including two very strong signals that spanned the regions of genes *six6* and *vgll3* on chromosomes 9 and 25, and this study for early maturation demonstrates that using very large sample sizes it is possible to reliably identify loci with small effect.

Regarding the molecular action of these variants on the physiology of maturation, there are several hypotheses related mainly to the *vgll3* gene. *Vgll3* is a regulator of adiposity in vertebrates [38]. Because the fat reserve level is considered a key element in the control of the initiation of maturation [39], its association with the age of maturation in Atlantic salmon seems clear and direct. The *vgll3* gene has two missense mutations strongly associated with age at maturation at amino acids 54 and 323 [14]. The haplotype associated with late maturation (e.g., 3 sea winters) codes for the amino acids threonine (Thr) and lysine (Lys), while those associated with early maturation (1 sea winter) code for methionine (Met) and aspartic acid (Asp). To date, it is unknown how the other markers, such as SNPs found in the *ndufs4*, *rora*,

Population/ Strain	Origin	Gene	Chr	Applied statistics	Method used	Reference
Wild	Norway	vgll3, akap11	Ssa25	Logit model	GWAS	[22]
		six6	Ssa09			
	Norway	vgll3, chmp2B, akap11	Ssa25	Cochran–Mantel–Haenszel test (CMH)	GWAS	[14]
	France	vgll3, akap11	Ssa25	Latent environmental threshold model (LETM)	Genotyping by sequencing and linkage map	[26]
Mowi	Canada	E2F4, MDH, PQLC2, PGRC1, SYAP1, FRA10AC1, PAPL, 14–3-3 beta/alpha	Ssa10, Ssa02, Ssa13, Ssa16, Ssa28, Ssa01, Ssa01	Mixed model and scoring tests	GWAS	[36]
Mowi	Norway	vgll3, mst1, nf2, tead3, ywhab, taz	—	—	qPCR	[14]
AquaGen					Analysis of global gene expression with RNAseq	[27]
Saint John-SJR	Canada	ropn1	Ssa21	Mixed model and principal components analysis	GWAS	[35]
SALTAS	Tasma-nia	lefty2, adpgk, scrib, ppp1r9a, pou6f2, cfap299, slitrk6, smarcd3, maggi2, picalm, ctnna2, rapgef2, cldn4, caln1, tsku, ser1, nlrp12	Ssa9, Ssa29, Ssa14, Ssa24, Ssa17, Ssa10, Ssa11, Ssa09, Ssa04, Ssa05, Ssa01	Mixed linear method	GWAS	[6]
Neva River	Finland	arhgap6e, akap11a, yap1 rd3l	—	—	NanoString transcription profiling	[37]
AquaGen	Norway	ndufs4, rora, cntn4	Ssa15, Ssa16, Ssa22	Mixed linear model	GWAS	[34]

Table 1.
Genes associated with age at maturation in wild and domesticated populations of Salmo salar.

cntn4 genes [34], influence the process of sexual maturation. Therefore, it is essential to continue functional genetic research to fill this knowledge gap in the future.

6. Conclusions

Early maturation is a serious problem for fish farms due to the economic losses, especially in periods close to harvest. Faced with this problem, various control strategies have been used, with photoperiod manipulation on sea cage being widely used in Atlantic salmon. In addition, triggered by fast development of the field of genomics, and based on growing knowledge on genetic components of sexual maturation, the genetic control of maturation is actively being implemented in selective breeding programs. The integration of both strategies should allow progress towards an effective control of early maturation in salmon aquaculture, both in marine aquaculture and in land-based aquaculture.

Acknowledgements

This work was funded by the National Agency for Research and Development (ANID)/Scholarship Program/DOCTORADO BECAS CHILE/2021 – 21211159. This research study also was funded by CONICYT-Chile through project FONDECYT N°1140772 and by the Cooperative Research Program Fellowships of OECD (PCI 2015-CONICYT) awarded to J.A.G.

Conflict of interest

The authors declare that they have no conflicts of interest.

Author details

Patricia Rivera[1], José Gallardo[1]*, Cristian Araneda[2] and Anti Vasemägi[3]

1 Laboratory of Genetics and Genomics Applied of Aquaculture, Pontificia Universidad Católica de Valparaíso (Pontifical Catholic University of Valparaíso), Chile

2 Faculty of Agronomic Sciences, Department of Animal Production, Laboratory of Genetics and Biotechnology in Aquaculture, University of Chile, Santiago, Chile

3 Department of Aquatic Resources, Swedish University of Agricultural Sciences, Drottningholm, Sweden

*Address all correspondence to: jose.gallardo@pucv.cl

References

[1] McClure CA, Hammell KL, Moore M, Dohoo IR, Burnley H. Risk factors for early sexual maturation in Atlantic salmon in seawater farms in New Brunswick and Nova Scotia, Canada. Aquaculture. 2007 Nov;272(1-4):370-379. DOI:10.1016/j.aquaculture.2007.08.039

[2] Good C, Davidson J. A review of factors influencing maturation of Atlantic Salmon, Salmo salar, with focus on water recirculation aquaculture system environments. J World Aquac Soc. 2016 Oct 9;47(5):605-632. DOI:10.1111/jwas.12342

[3] Carillo M, Zanuy S, Bayarri M. El control ambiental de la reproduccion de los peces con especial referencia al control del ciclo sexual, de la pubertad y de la precocidad. In: Espinosa de los Monteros J, editor. La reproducción de los peces: aspectos básicos y sus aplicaciones en acuicultura. Madrid: Pub. Cient. Tec. FOESA – CSIC; 2009. p. 175-246. Available from: Fundacion Observatorio Español de Acuicultura

[4] Fleming IA, Reynolds JD. Salmonid Breeding System. Evol Illum – Salmon their Relat. 2004;(January 2004): 264-294.

[5] Webb J, Verspoor E, Aubin-Horth N, Romakkaniemi A, Amiro P. The Atlantic Salmon. In: Verspoor E, Stradmeyer L, Nielsen J, editors. The Atlantic Salmon: Genetics, Conservation and Management. Blackwell Publishing Ltd; 2007. p. 17-56. DOI:10.1002/9780470995846.ch2

[6] Mohamed AR, Verbyla KL, Al-Mamun HA, McWilliam S, Evans B, King H, et al. Polygenic and sex specific architecture for two maturation traits in farmed Atlantic salmon. BMC Genomics. 2019 Dec 15;20(1):139. DOI:10.1186/s12864-019-5525-4

[7] Guerrero-Tortolero DA, Bromage N. Growth and maturation of Atlantic salmon (Salmo salar) populations with different grilse proportions under natural photoperiod and superimposed nighttime light. Aquaculture. 2008;285(1-4):63-66. DOI:10.1016/j.aquaculture.2008.07.045

[8] Whalen KG, Parrish DL. Effect of maturation on parr growth and smolt recruitment of Atlantic salmon. Can J Fish Aquat Sci. 1999 Jan 1;56(1):79-86. DOI:10.1139/f98-154

[9] Taranger GL, Carrillo M, Schulz RW, Fontaine P, Zanuy S, Felip A, et al. Control of puberty in farmed fish. Gen Comp Endocrinol. 2010;165(3):483-515. DOI:10.1016/j.ygcen.2009.05.004

[10] Jonsson B, Jonsson N. A review of the likely effects of climate change on anadromous Atlantic salmon Salmo salar and brown trout Salmo trutta, with particular reference to water temperature and flow. J Fish Biol. 2009 Dec;75(10):2381-2447. DOI:10.1111/j.1095-8649.2009.02380.x

[11] Thorstad E, Whoriskey F, Rikardsen A, Aarestrup K. Aquatic nomads: The life and migrations of the Atlantic Salmon. In: Aas Ø, Einum S, Klemetsen A, Skurdal J, editors. Atlantic Salmon Ecology. Blackwell Publishing Ltd; 2011. p. 1-23.

[12] Vähä J-P, Erkinaro J, Niemela E, Primmer CR. Life-history and habitat features influence the within-river genetic structure of Atlantic salmon. Mol Ecol. 2007 Jul;16(13):2638-2654. DOI:10.1111/j.1365-294X.2007.03329.x

[13] Fjelldal PG, Schulz R, Nilsen TO, Andersson E, Norberg B, Hansen TJ. Sexual maturation and smoltification in domesticated Atlantic salmon (Salmo salar L.) – Is there a developmental conflict? Physiol Rep. 2018 Sep;6(17): e13809. DOI:10.14814/phy2.13809

[14] Ayllon F, Kjærner-Semb E, Furmanek T, Wennevik V, Solberg MF, Dahle G, et al. The vgll3 Locus Controls Age at Maturity in Wild and Domesticated Atlantic Salmon (*Salmo salar* L.) Males. Houston R, editor. PLOS Genetic. 2015 Nov 9;11(11):e1005628. DOI:10.1371/journal.pgen.1005628

[15] Peterson RH, Harmon PR. Changes in condition factor and gonadosomatic index in maturing and non-maturing Atlantic salmon (Salmo salar L.) in bay of Fundy Sea cages, and the effectiveness of photoperiod manipulation in reducing early maturation. Aquac Res. 2005 Jun;36(9):882-889. DOI:10.1111/j. 1365-2109.2005.01297.x

[16] Iversen M, Myhr AI, Wargelius A. Approaches for delaying sexual maturation in salmon and their possible ecological and ethical implications. J Appl Aquac. 2016;28(4):330-369. DOI:1 0.1080/10454438.2016.1212756

[17] Thomson AL. Factors determining the breeding seasons of birds: An introductory review. Ibis (Lond 1859). 2008 Apr 3;92(2):173-184. DOI:10.1111/ j.1474-919X.1950.tb01748.x

[18] Crampton WGR, Rodríguez-Cattáneo A, Lovejoy NR, Caputi AA. Proximate and ultimate causes of signal diversity in the electric fish Gymnotus. Krahe R, fortune E, editors. J Exp Biol. 2013 Jul 1;216(13):2523-2541. DOI:10.1242/jeb.083261

[19] Qiu D, Xu S, Song C, Chi L, Li X, Sun G, et al. Effects of spectral composition, photoperiod and light intensity on the gonadal development of Atlantic salmon Salmo salar in recirculating aquaculture systems (RAS). Chinese J Oceanol Limnol. 2015;33(1):45-56. DOI:10.1007/ s00343-015-4011-3

[20] Taranger GL, Hansen T. Ovulation and egg survival following exposure of Atlantic salmon, Salmo salar L., broodstock to different water temperatures. Aquac Res. 1993;24(2):151-156. DOI:10.1111/j. 1365-2109.1993.tb00535.x

[21] Fleming IA. Pattern and variability in the breeding system of Atlantic salmon (Salmo salar), with comparisons to other salmonids. Can J Fish Aquat Sci. 1998;55(S1):59-76. DOI:10.1139/ cjfas-55-s1-59

[22] Barson NJ, Aykanat T, Hindar K, Baranski M, Bolstad GH, Fiske P, et al. Sex-dependent dominance at a single locus maintains variation in age at maturity in salmon. Nature. 2015;528(7582):405-408. DOI:10.1038/ nature16062

[23] Muñoz-Cueto JA. Cerebro y reproducción en peces: bases neurales y neuroendocrinas. In: Carrillo M, editor. La reproducción de los peces: aspectos básicos y sus aplicaciones en acuicultura. Fundación observatorio español de Acuicultura; 2009. p. 25-71.

[24] Falcón J, Migaud H, Muñoz-Cueto JA, Carrillo M. Current knowledge on the melatonin system in teleost fish. Gen Comp Endocrinol. 2010;165(3):469-482. DOI:10.1016/j. ygcen.2009.04.026

[25] Verta J, Debes PV, Piavchenko N, Ruokolainen A, Moustakas-verho JE. Regulatory divergence in vgll3 underlies variation in age at maturity in male Atlantic salmon. bioRxiv. 2019;1-20.

[26] Lepais O, Manicki A, Glise S, Buoro M, Bardonnet A. Genetic architecture of threshold reaction norms for male alternative reproductive tactics in Atlantic salmon (Salmo salar L.). Sci Rep. 2017;7(January):1-13. DOI:10.1038/ srep43552

[27] Kjærner-Semb E, Ayllon F, Kleppe L, Sørhus E, Skaftnesmo K, Furmanek T, et al. Vgll3 and the hippo

pathway are regulated in Sertoli cells upon entry and during puberty in Atlantic salmon testis. Sci Rep. 2018;8(1):1-11. DOI:10.1038/s41598-018-20308-1

[28] Ayllon F, Solberg MF, Glover KA, Mohammadi F, Kjærner-Semb E, Fjelldal PG, et al. The influence of vgll3 genotypes on sea age at maturity is altered in farmed mowi strain Atlantic salmon. BMC Genet. 2019 Dec 6;20(1):44. DOI:10.1186/s12863-019-0745-9

[29] Taranger GL, Haux C, Stefansson SO, Björn Thrandur Björnsson, Walther BT, Hansen T. abrupt changes in photoperiod affect age at maturity, timing of ovulation and plasma testosterone and oestradiol-17β profiles in Aatlantic salmon, Salmo salar. Aquaculture. 1998;162(1-2):85-98. DOI:10.1016/S0044-8486(98)00168-9

[30] Hansen TJ, Fjelldal PG, Folkedal O, Vågseth T, Oppedal F. Effects of light source and intensity on sexual maturation, growth and swimming behaviour of Atlantic salmon in sea cages. Aquac Environ Interact. 2017;9:193-204. DOI:10.3354/AEI00224

[31] Good C, Weber GM, May T, Davidson J, Summerfelt S. Reduced photoperiod (18 h light vs. 24 h light) during first-year rearing associated with increased early male maturation in Atlantic salmon Salmo salar cultured in a freshwater recirculation aquaculture system. Aquac Res. 2016;47(9):3023-3027. DOI:10.1111/are.12741

[32] Gjerde B. Response to individual selection for age at sexual maturity in Atlantic salmon. Aquaculture. 1984;38(3):229-240. DOI:10.1016/0044-8486(84)90147-9

[33] Lhorente JP, Araneda M, Neira R, Yáñez JM. Advances in genetic improvement for salmon and trout aquaculture: The Chilean situation and prospects. Rev Aquac. 2019;11(2):340-353. DOI:10.1016/0044-8486(84)90147-9

[34] Sinclair-Waters M, Ødegård J, Korsvoll SA, Moen T, Lien S, Primmer CR, et al. Beyond large-effect loci: Large-scale GWAS reveals a mixed large-effect and polygenic architecture for age at maturity of Atlantic salmon. Genet Sel Evol [Internet]. 2020 Dec 12;52(1):9. DOI:10.1186/s12711-020-0529-8

[35] Boulding EG, Ang KP, Elliott JAK, Frank P, Schaefferc LR. Differences in genetic architecture between continents at a major locus previously associated with sea age at sexual maturity in European Atlantic salmon. Aquaculture [Internet]. 2019;500:670-678. DOI:10.1016/j.aquaculture.2018.09.025

[36] Gutierrez AP, Yáñ Ez JM, Fukui S, Swift B, Davidson WS. Genome-wide association study (GWAS) for growth rate and age at sexual maturation in Atlantic salmon (Salmo salar). PLoS One [Internet]. 2015;10(3):1-15. DOI:10.1371/journal.pone.0119730

[37] Kurko J, Debes P V., House A, Aykanat T, Erkinaro J, Primmer CR. Transcription profiles of age-at-maturity-associated genes suggest cell fate commitment regulation as a key factor in the Atlantic salmon maturation. bioRxiv [Internet]. 2019;10(January):235-246. DOI:10.1101/778498

[38] Halperin DS, Pan C, Lusis AJ, Tontonoz P. Vestigial-like 3 is an inhibitor of adipocyte differentiation. J Lipid Res [Internet]. 2013;54(2):473-481. DOI:10.1194/jlr.M032755

[39] Thorpe JE, Mangel M, Metcalfe NB, Huntingford FA. Modelling the proximate basis of salmonid life-history variation, with application to Atlantic salmon, Salmo salar L. Evol Ecol [Internet]. 1998;12(5):581-599. DOI:10.1023/A:1022351814644